About This Book

- Spanning almost ¾ of a century, and covering a gazillion square miles of Canada and the United States, it is a study/commentary of how one can become a rolling stone, gathering little moss.

- But – rolling stones can, and do, gather a wealth of experience and insight.

- During the construction of Books One and Two the authors attempted to downplay the "we/me" bit with some success. However – this book is more of the "I/we/us/" type and it had to be so.

- If you are a generation or two younger than the authors, keep your own memories in your treasure-chest. You will also enjoy taking them out occasionally to study on.

2

I Call Myself a Prospector

Book Three

Fiddle Foot Training

By

Bob Durnin

With

Frank Durnin

Format and Design by Coreshack Publishing

(Available at Amazon.com)

Acknowledgements

We thank:

- Our two remaining siblings. Some early events pre-dated us, and although there were some conflicting points of view, we feel that by and large, accuracy prevails.

- Our memory safety filters – some stuff remains unexhumed.

- Our Mother's scrapbook – especially the Mumbo-Jumbo poem. We feel this is the highlight of Book Three.

- The Ethernet – who knew that pics of old cars could so well illustrate our youth?

- And thanks to Stan and Irene Olson for the great pics and the wolf story. Is it true? I didn't dare ask Stan. But once again – it is a highlite.

- Last but not least – credits must be bestowed upon my co-author. We had our disagreements, but have not yet come to blows. The beauty of working six hundred miles apart – beyond pistol range.

"Memories fade – they must be preserved."

– Peter C. Newman

(Company of Adventurers, Vol 1)

Foreword

A friend was given an early draft of Book One to read and to offer an opinion, which he did. Ignoring the poor grammar, sentence structure and abysmal hodgepodge of semi- connected stories, he had only one input. "How did you happen to become such a fiddle-foot"?

Point well taken. Thank you Bob Marvin.

So I took my pen in hand, my brother clipped the nails on his two forefingers and thus we have:

BOOK 3 - FIDDLE FOOT TRAINING

Prologue

Every story should start at the beginning. This one starts before I appeared on the scene.

Mom and Dad, New Brunswick.

Table of Contents

Fiddle Foot Training

Chapter I

My father was born in the twilight years of the nineteenth century near Olga, North Dakota. His parents had been married in Goderich, Ontario and had headed west for free prairie land. Grandpa hedged his bets, homesteading in N.D. under one variation of his given names and in Manitou, Manitoba under another. Dad's mother died shortly after he turned three.

Grandpa remarried and Dad was not treated well by his step-mother although his step-siblings were kinder. At 14 he pulled out, hit the harvest trail and spent a few years working at various farms. He must have had some tough times but never, to my knowledge did he speak of them. Most of his stories were of good memories – some were very interesting.

He told about plowing with huge steam tractors pulling a 12 bottom (individual moldboards) plow. A narrow platform ran along the leading edge of the plow and at the end of the field it was Dad's job to trot along the platform lifting each moldboard in turn, thus leaving the head land straight and even. Then he would jump onto the steam tractor to help the driver crank the steering wheel, then back onto the platform to drop the plows one at a time.

The large steamers had no automotive-type steering. A chain ran from one front axle end to a cog drum and back to the other end of the axle which had a centre pivot point. When the men cranked the steering wheel one chain shortened and the other lengthened – "Armstrong Power Steering."

He also spoke of driving up to six horses on a cultivator or a set of harrows. Dad was quite a horseman in his day.

One of Dad's earliest memories was of a fire in Thief River Falls, Minnesota. Every winter, farmers from N.D. and southern Mb. would take a team to Thief River Falls and haul logs to a sawmill. One night the horse barn burned to the ground with a loss of 300 horses! Even allowing for youthful exaggeration, I'm sure many, many horses died in the fire – it was a huge mill.

(Drive by Thief River Falls nowadays. The only large trees are in farmyards and the nearest bush is miles to the east).

Sidebar: There were (and still are) more than a few Durnins around Goderich, Ont. and another branch of the family had settled near Brandon, Manitoba. In the early 80's I was on a pipeline inspection tour with Herbie Schmidt. We were stopped at an intersection on the outskirts of Brandon and a gravel truck drove by. On the door was lettered "ROBERT DURNIN TRUCKING".

"Run that guy down, Herbie," I said, "He stole my truck."

When WWI broke out our dad volunteered and was a gunner in a howitzer battalion. His service record shows that he hit almost all the bad ones. During this time he received a battlefield commission, and we all know what that means – no officers left alive. He would never join the Canadian Legion and very, very seldom talked about the war itself, preferring side-lights such as pulling the howitzers up the cliffs into France with 9 span (pairs) of mules!

One day I screwed up enough courage to ask him how it was over there. He told me that surely the trenches were hell, but the big guns were no pieces of cake either. Each side was trying to take the other out. One day an enemy shell hit within 20 feet of his gun emplacement showering them with mud. The mud was so deep that the shell did not detonate. Had it done so we would not be here to tell these tales.

"Jerry never found the range again," he said.

Along with his uniform, medals and such there is a letter from his favorite cousin Earl, who was killed elsewhere in France. Dad never mentioned Earl, but the fact that his last letter was kept all those years indicated the depth of Dad's loss.

Frank's Take

One thing we learned was that dad "spotted" for his mates, an exceedingly dangerous position. Spotters were exposed, and enemy snipers could and would pick them off.

How ironic – Dad – in the most dangerous job, probably escaped being wiped out when an incoming shell did its job.

(The following is a copy of his WWI service record showing his promotion to Lieutenant and his return to action on April 5 1918 in charge of his own gun.)

Sgt in
184 Batt
Dec 1 1916
transferred
July 4 1916
gunner
76th Depot
Battery
C F A
promoted
Lieutenant
April 5
1918

Vimy Ridge
Hill 70
Passchendaele
Arras
Cambrai
Mons
Returned to England
April 1 1919
Demobilized
Winnipeg M.D. 10
May 13. 1919

not applicable   not applicable   not applicable   not applicable   not applicable

Counstersigned by .... (Signature of Senior Officer).
.... (Rank and Regiment).
.... (Date).
or Certified that this Return is correct.

Certified as correct.
.... (Signature of Officer).
Forwarded.

Date Feb 2nd 1920

Sidebar: In 1965 I was a social member of a Legion and got to know two brothers, both WWII vets. One, when in his cups, fought the war heroically. I later found out that he never left England.

The other brother, a very nice mild mannered gent, was very pleasant to talk to about much interesting stuff, but never, never did he discuss his war time experiences.

He was the most highly decorated of that Legion's membership.

After the war Dad taught school in the Interlake District in Manitoba. This was where he met Mom.

She was born in 1900 at West Glassville, New Brunswick, a farming area just over the height of land, 5 miles east of the St. John River. The nearest town was Bath, situated on the river, and not too many miles north of Hartland, home of the longest covered bridge in the world. (The original structure burned a number of years ago. It was rebuilt and is now only used for foot and cycle traffic. I crossed it a number of times when I was an ankle-biter.)

She was a descendent of United Empire Loyalist stock. Her mother's maiden name was Rogers. Her great, great grandfather Rogers rode with De Lancey's Raiders on the British side during the American Revolution and, in fact, they occupied New York City during part of one winter. (If King George III had shown any interest in his Canadian colony other than as a beaver top-hat generator, the Treaty of Ghent might have resulted in Vermont and Maine being part of Canada.)

Mom's younger years were similar to Dad's. She was the youngest of 12 surviving children. One sister died young and the last baby was still-born, and her mother died shortly after.

Grandpa married a widow with children of her own and as happened to Dad, the step-mother treated Mom and those who remained at home badly. Mom said that she got bread and lard sandwiches for lunch, sometimes with butter, and even more rarely with jam. Meanwhile, the step-siblings enjoyed roast beef, roast pork and chicken for lunch.

Those things happened 100 years ago and still do, but times could be more harsh back then, as in the case of Mom's two older brothers, who in their teens fell victim to the dreaded tuberculosis.

At that time there was no magical medical wand. The conventional treatment was rest and isolation, so with no sanatorium available to go to, the boys were put in a tent in the back forty. Mom and an older sister delivered food to them, usually oatmeal porridge. They had to set the food on the ground and back away while the boys ate – no kisses, hugs or cuddles – and it was so difficult for the girls. One of the boys did not make it. The younger one beat the TB and moved to Alberta only to die in a farm accident before he was 20.

But families stuck together. Mom's oldest sister took her out of that situation, made sure she finished high school and backed her for nursing training in Presque Isle, Maine. Mom became a Registered Nurse and that stood us all in good stead in the years to come.

Another sister took Thelma, (my Sudbury Aunt) and made sure she received training as a classical pianist. Another sister became an actress and was mysteriously rubbed out in Ohio. Rumor has it that she was shot by a jealous husband. A very veiled rumor – no one talked much about Aunt Rose.

Sidebar: Post-1812 when Canada and the US of A decided to stop burning down each other's important buildings, the dotted line between our two countries was pretty much just dots. Thus Mom was able to train as a nurse in Maine, Aunt Vivie married an investment banker in New York and Aunt Rose was allowed to run afoul of a jealous husband in Ohio - no green card or immigration hassles. As for the Manitoba/N. Dakota Border, I doubt that the wagon had to clear customs when Grandpa, Grandma and Dad went to Thief River Falls every winter. Likewise, our Rainy River, which separates us from Minnesota, had no watch towers or fences. One could live on side and work on the other – no big hairy deal.

Table Scrap

When the wife and I operated our restaurant in Mine Centre during 1988-90 we got to know Lyle Laidlaw, a nice old gent from International Falls, Minn. He stopped in 2 or 3 times a year and I got to know him quite well. He had an interesting back story.

Lyle was born around 1920 on a farm two miles south of Devlin (the town where we would open our C-store/gas bar/restaurant in late 1990). In his teens he hired on at the paper mill in Fort Frances as an electrical apprentice and spent a lot of time at the two power dams on the Seine River, 20 miles east of Mine Centre. At that time only the railroad existed, so Lyle would have stayed on site most of the time. When WWII broke out Lyle enlisted and went overseas. On his return he got his job back and completed his apprenticeship in the mill on the American side.

Married now, he and his wife lived in Fort Frances. Lyle crossed the bridge every day to go to work. At that time one had to check in at both customs entry points, back and forth, forth and back.

Lyle and his wife had a routine. Every second Friday (pay day) she would join him in the Falls after work. They would dine out and spend the evening socializing at the V.F.W. Club (Veterans of Foreign Wars) or the American Legion, returning home to the Fort in the wee hours.

One evening in 1948 they were walking home and at the American customs an officer took Lyle aside and told him to let his wife go home alone. Lyle asked why and the officer said "Never mind why. Go back, get a room at the Rex Hotel and see us in the morning. Lyle did so.

Saturday morning the border shutdown. If one woke up in the US of A one was a landed immigrant - no hoops to jump through. Had Lyle gone home that night he would not have been allowed to keep working at the Falls mill without a green card.

It was no problem for his wife of course. They moved to the American side and Lyle retired as electrical superintendent in 1985.

Table Scrap: Lyle-Part Two

This deals with my favorite bete noir – Bureaucratic Baffle Gab.

Lyle tracked me down again when we added the restaurant to the store in Devlin. He was alone now. His wife had passed on and they had no children. We talked about stuff and he was very interested in all things local. He seemed to want to re-connect to his youth.

I told him we had a new school a mile up the road. It was K to grade eight and although it was a spiffy deal with a sports field, there had been no money left in the budget for playground equipment for the little ones.

I told him that a group of younger mothers had formed "Playground Parents" and were attempting to raise money to buy equipment. We were doing our part. We had a popcorn machine and from time to time we would pop a bushel or two and the ladies would package popcorn in little bags which they would sell to the rugrats at 10 cents a bag. The kids were essentially subsidizing their own playground. Lyle perked up – he was very interested.

He asked me to set it up for him. He would like to donate $2000 to the project, Wow!

So I checked with the ladies and they also said, "Wow!" They were nervous about how to handle such a large gift, and checked with the district school board, something they should have done much earlier (or not done at all.)

A bureaucrat or two gleefully (I'm sure) informed the head lady that ordinary, unwashed citizens could not, in fact, purchase, supply or deliver unauthorized playground equipment to a school owned and funded by taxation derived from the common people. Funds were to go to the Central School Board where elected (sanitized) members would decide how and where the money would go.

Lyle and I were disappointed and disgusted, and he kept his two grand in his bank account south of the border.

After our fire I lost track of Lyle. I assume he joined his wife. I also assume that some of his stash went to his adopted government and I hope some of it was used to buy "Made in the USA" playground stuff.

Table scrap: Tough Times make Tough People

Take the case of Great Aunt Hepzibah (aka Hepsie). This one was told to me by cousin Charlie, our family historian who dated the tale in mid-to-late 1800s.

Hepsie and two children were alone in their cabin in the New Brunswick woods. She and her husband had cleared enough land for a garden and a 10-acre field, but ready cash was hard to come by, so her husband had taken the horses to a lumber camp and would be away 'til spring.

He had left her in good shape. Potatoes and carrots and such were in the root cellar along with canned veggies and wild fruit preserves. There was hay in the loft of the little barn for the cow and her calf and a well in the front yard. With salt pork, bacon and staples on hand and two sides of venison hanging in the well-stocked attached wood shed they were set for the winter.

But then a mountain lion moved into the woodshed! The next morning Hepsie went to enter the shed through the interconnecting door off the kitchen and was greeted by a deep growl! She jumped back and dropped the wooden security bar into place.

Later that day she saw the cougar prowling in the yard and made another attempt to fetch wood. The cat headed for the open front of the shed and Hepsie jumped back into the kitchen. This time he stayed put – the meat was his and he was guarding it.

On the second day Hepsie was out of water but managed to scoop up some snow beside the front door. She had been stingy with the wood box but when the wood ran out she could no longer melt snow for drinking or cooking. Handfulls of snow kept her and the children from becoming seriously dehydrated, but the cow and calf had been bawling since yesterday.

When the first sun hit the woodshed on day three, so did Aunt Hepsie, axe in hand. The mountain lion didn't stand a chance and I'm sure his last thought was "Maybe I should have quit when I was ahead."

Table Scrap: The Old Gray Mare

This one is from is from our Mom's old scrapbook and although the scrapbook was started at least 100 years ago we feel this story may also date back to the previous century.

You ask me if that horse is for sale, Sir. Well, she is not, nor will she ever be sold. That Betsy girl saved our lives a good many winters ago.

It was a hard winter that year, snowing almost every day until there was more than four feet in the woods. Our hired man, John, and I had been yarding out timber and delivering logs to the sawmill in town, so the ten-mile road was well packed. With no hills to speak of, we had no problem hauling good-sized sleigh-loads to the mill. By mid-January the weather turned colder and it stopped snowing. The road was now as smooth and strong as a turnpike.

One Saturday we hitched Betsy to the cutter and the wife and I and our two little ones headed to the village for supplies. It was a pleasant trip in the January sun, with the runners creaking on the hard snow and the little mare trotting along quite easily.

In town the folks were talking about a wolf pack that had been taking some livestock. The deep snow had made wild game difficult to bring down and the local farmers were guarding their herds well. We were unworried – wolves had never been a big problem for us.

What with visiting and such we were late leaving for home. By the time we passed the last farmhouse two miles out, the sun had set, a full moon had risen, and we and Betsy could see the road well.

She sensed the danger before we did. Her head was up, ears scanning, and her pace picked up to a brisk trot.

Now we could hear it, too – howling off into the woods, getting closer as we moved along. With five miles to go the howls abated somewhat, but a glance over my shoulder told me why. The pack had come out of the woods behind us and were hot on our trail. They smelled fresh meat!

Betsy was galloping now, she needed no urging and I gave her free rein. The wolves had stopped howling, but they were gaining ground, and soon we could hear snarls and yips behind us so we started jettisoning cargo – not much there to interest a hungry wolf – flour, sugar and such, but it brought us some time and lightened the load.

Two miles to go and Betsy was running flat out, belly down, head outstretched and ears folded back. Clods of packed snow from her hooves were whistling past our ears. A big lobo came up beside the sleigh and I dropped it with my rifle. This brought us a little more time, but with half a mile to go the pack was on our tail again.

John heard us coming and he had the barn doors open. In we went and the doors closed behind us. One wolf made it through and John dispatched it with his axe.

That hell-bent-for-leather run in the cold air didn't do Betsy's lungs any good and she never worked again, but No, Sir – she is <u>not</u> for sale!

Table Scrap: Another one from Charlie Jones' Memory Locker

It's 1905. Grandpa Jones, Uncle Guy (the oldest boy), and Uncle Walter (Charlie's Dad) are cutting birch timbers on the farm at West Glassville. The equipment used is basic stuff: a team of horses, a wooden wheeled wagon, a couple of New Brunswick double-bitted axes and an adze. The wagon has no rack, bolsters with short pegs at each end will hold the timbers. The double-bit axes have non directional handles and are heavy. The adze, also heavy and shaped like an old English broad-axe is used to square up the timbers.

The wood lot is virgin forest. The birches are big, tall and straight, with no limbs for the first 40 feet. One tree will fall today and be divided into four 10-foot timbers. Two will be 24" square, two will be 20" to a side.

They drop the first tree and while Guy cuts off the first 20' foot section Grandpa and Walter start squaring it up. (Charlie was told that when Uncle Guy finished the cut you would swear it was sawn – Guy was an axeman for sure).

With all timbers squared they skid them out to the wagon and load them. These logs are heavy and the men use a ramp. Two smaller logs are leaned against the wagon wheels. Holes have been augured on one side of these poles and pegs fit therein. One end of the logs is moved up the ramp and secured by a peg. Then the other end is slid up likewise and by working back and forth the timber reaches the bolsters.

While Grandpa and Guy fell another birch, Walter takes the load 5 miles to the height of land above the village of Bath. It is fairly gentle upward slope but the timbers are heavy and give the horses a workout. At the top of the hill he rolls the logs off the wagon and returns to the wood lot where Grandpa and Guy have the second load ready.

They return to the hill, reload the timbers Walter had dumped, and now they are faced with a steep downgrade to the village. With the team leaning back in the harness breeches and with one man on each rear wheel brake they arrive safely at level ground above the river. There is a steep bank, but they merely roll the logs onto a sluiceway where they slide down to join other timbers floating inside a log boom. The timber buyer will take the boom down-river to a place where they will be loaded as ballast for ships returning to England. When they reach England they will be re-sawn into lumber for building materials or for good strong knot-free furniture.

The crew returns home, stopping at the wood lot to pick up tops and branches for next winter's firewood. Even the chips are kept for the smokehouse.

At home a late supper is ready. The younger children have milked the cows, slopped the hogs and the chickens are on their roosts. After eating they sit on the veranda with a cup of tea and a pipe to watch the sun set in the west. It has been a long, hard prosperous day and tomorrow two more giant birch will fall to the axe.

Grandpa had been paid $20 for the timbers. He keeps $10 and gives the boys $5 each. They will share with the younger kids – perhaps a new pinafore or a warm shirt for school in September.

Subscrap: Golden Birch

In 1962 I worked with a guy who shared this story.

His mother was from the Dokis Indian Reserve near Sturgeon Falls Ontario. On the reserve were acres of old-growth birch, harvested in the past to build birch-bark canoes.

Nowadays, they select-cut a few every year. These went to a mill to become veneer supplying the exterior layer for cabinetry plywood. Each individual tree brought in up to

$4000 1962 dollars and the cash was shared pro rata within the reserve.

Sub-subscrap: Silver Birch – Source: Kerr-Addison Jim (Book Two)

Jim had a crew staking claims near a shoreline on Lake of the Woods. One line led into the lake through a cottage lot. The staker needed a "witness post" and, believe it or not, used a nice, straight, six-inch silver birch in the cottager's front yard.

The forensic investigation was not difficult – the staker's name was on the post. That not-yet fully mature birch cost Kerr-Addison $800 and cost the staker his job.

Corollary: Whether lining a canoe's exterior, pockets of West Glassville pants, or shorelines, birches are and have always been a valuable commodity.

With her prized R.N. certificate in hand Mom worked as a personal nurse for the McCain Family, tending to Mrs. McCain. With her patient on the road to recovery she headed west to look for work. Two sisters were living in the North Interlake District of Manitoba and she stopped to visit them. Here she met Dad who was teaching at a one room school east of Ashern and sparks flew. Mom went on to Saskatoon to work in a hospital. Dad followed her and went to work as a mail carrier and shortly thereafter they were married. After the birth of my oldest sister they went back to the land. Farming was in our dad's blood.

The first attempt was with his Uncle Rob (our Great Uncle.) Rob and his wife had no children and were very successful farmers, but is as often the case with free labor building partnership equity, things happen. The deal imploded after the birth of my older brother and Mom, Dad and the two children moved to a rented farm at Elstow, near Saskatoon.

Things went well for a time. They bought some machinery, did some cash cropping and put together a small dairy herd. Daily milk deliveries to the village brought in a regular income – and then the kids got the measles.

Even now an outbreak of measles can cause a public kerfuffle, but in those days the 3 P's came into effect. People were Positively Paranoid.

Mom, being a nurse, would have recognized the early signs and Dad immediately moved to the barn. Also, we assume that Mom would have felt duty bound to report the measles to the Department of Health. A large red quarantine sign was nailed to the front door and in two shakes of a lamb's tail the milk business dried up. The creamery refused to accept the cream and with no income they had to pack it in. They sold what they could (would anyone buy the cows?) and taking some machinery and horses with them they homesteaded at Maidstone, Saskatchewan, a few miles from Lloydminster.

It was 1934. Could they have chosen a worse time to homestead? The stock market had crashed in 29 and the dirty thirties were in full non-bloom. They built a log house and barely survived.

This is the log house at Maidstone. Mom and Dad built it, the older kids stomped straw into mud to make plaster for the walls. It held out the western winter quite well – dirt floor and all. (Mom's visit in 1977)

Sidebar: Mom did some midwifery in the thirties. This may have brought in some badly needed cash, but more often a chicken or a weaned piglet was all a family could afford.

In 1977 my co-author took Mom to Maidstone's 60th anniversary celebration in the community hall. Mom was invited to sit at the head table and the Master of Ceremonies posed two questions to the audience.

"How many of you were brought into this world by a doctor"? Four stood up.

"How many were assisted by Mrs. Durnin"? All over the hall folks stood, and Mom got a standing ovation. She was pretty proud.

By 1938 Mom was tired. Tired of crop failures, hungry children and endless dust. She told Dad she was going back to New Brunswick and that was not negotiable. So they sold the livestock for pennies per dollar's worth, loaded up the Model T and headed east, stopping off in the Manitoba Interlake, where two sisters were also pulling out. One had been farming on her own for a year, the depression having driven her husband certifiably mad.

Two cars left for New Brunswick. Aunt Blanche and Uncle Chole drove their Gray-Dort (a Canadian automobile manufactured in Chatham, Ontario.) Their son John drove the Model T. and Dad stayed behind to dispose of the other aunt's chattels, such as they were. He would join them later.

They must have travelled through the northern states, as the Trans-Canada Hwy would not be completed until 1940. There was not much room in the cars. The youngest boy, Rodney, rode the entire trip sitting on a square 10 lb. Salada Tea container. (He has since told me that the tea canister hung around the house for a few years after that trip and you could still see the imprint of his tiny bum on the lid.)

The family linens, heirlooms and such had been shipped by rail in two steamer trunks. Only one arrived. The other was never to be seen again.

Table Scrap

How do I know that the Trans-Canada (Ontario) was completed in 1940? I didn't have to Google this one – my old friend Sam Duggan (Book Two) told me. You will recall that I spent a lot of time with Sam in 1996 before cancer did him in.

Sam was raised on a farm in Manitoba near Portage La Prairie. Even as a youngster Sam was a charmer and at 16 he caused a school board kerfuffle pertaining to a young school marm resulting in her termination. His father suggested very strongly that it was time for Sam to find his own path in life.

So Sam headed east by rail to Geraldton, Ontario and went to work in one of the gold mines. It was a good fit for him and Sam would be involved in hard rock mining and exploration for the rest of his life.

In 1939 Sam was eighteen and he bought a new, loaded 1939 Chev Master Deluxe four-door. (Loaded in 1939 meant exterior sun visor, radio, heater, bumper guards and whitewall tires.) In May 1940 he took some time off and headed home to show Dad his new car, rightfully proud of his success.

He knew the last connection was under construction somewhere between Beardmore and Nipigon, but when he got to the end of the gravel he was told that the spring rains had left the next 20 miles a sea of mud – the Chev would never make it. Sam told the spread boss his reason for heading west and the boss, who probably had sons of his own, whistled up a cat skinner.

Sam always said he was the first to drive the Trans-Canada – so what if it was on the end of a chain?

Table Scrap: Sam Again

(We'll ask our readers to cut us some slack, here. This tale is out of context, but we'll toss it into the mix anyhow. The thing is – one memory flash often triggers another.)

In the early 40's Sam was working underground at one of the Geraldton mines (theTom Bigbee?) Because the gold veins were fairly close together, they were using the room and pillar method. The stopes were large and pillars of rock were left to maintain some roof support. Sam said that one pillar at the head end of the stope obviously held a lot of high grade gold. All the shifts passed by that pillar and the gold was visible. When the stope played out, they planned to work back to the entrance taking out pillars as they retreated.

Sam was on the midnight shift and one night the shift boss pulled them off the back wall before the shift ended. The boss then told them to take out the high grade pillar, which they did, mucking the rock themselves and sending it up to the mill.

They went up in the personnel cage and Sam said he knew something was up – the day shift had not come down.

The underground workings were shut down when they reached the surface, with the mill and smelter following soon after. No one knew why – there was still gold at depth.

Perhaps the war was siphoning off manpower. Sam himself went home to Portage to join the RCAF. (I really don't know the mine's name for sure but it may have been, or is now being rehabilitated.)

Sub-scrap: Sam cont.

When the mine shut down Sam signed up with the RCAF planning to become a pilot, but he had some landing issues. One day the base commander called him in to discuss said issues. He suggested that Sam might be more valuable to the war effort as an airframe tech, putting it this way: "If you continue pilot training we will soon run out of planes."

Later on in life Sam held a private ticket and over the years would own and fly a 180 or two with only two landing incidents that I know of.

The first I'll not go into because, if you live in Red Lake or know anyone from Red Lake, even today, they will tell the tale. Let's just say it may have involved a drink or two, a glassy water landing and a 180 standing on its nose on a rip-rap berm protecting an av-gas storage tank on the shores of Howey Bay. Sam had some "'splainin'" to do on that one.

The second incident, Sam told me about in 1996.

He left Red Lake for Fort Frances one afternoon in good weather, and it being high summer, he had plenty of daylight for an evening arrival. However, he picked up a strong easterly crosswind which slowed him somewhat. Another problem arose – when he reached Rainy Lake a low overcast had come in and darkness descended earlier than anticipated.

He was landing in Sand Bay and Sand Bay was rolling. He made a pass into the wind and the waves were tremendous. Sam said he figured he would maybe get two bounces, but was sure that when he lost airspeed his floats would dig into a wave with an upside-down result. Daylight was virtually gone and there was no time to find a quieter lake.

So he landed with the wind. He could see the shoreline and the white-capped waves. He set the 180 down ever so gently on a big one and surfed in quite comfortably to Rusty Myers' dock.

In New Brunswick, as per usual, the family stayed with relatives. Dad, after disposing of Aunt Maude's stuff, hit the harvest in southern Manitoba to earn some cash before joining them. He wrote letters with money enclosed, and the letters were addressed to Bath, New Brunswick. Her Majesty's Royal Mail, efficient as always, sent them to Bath, Ontario.

It must have been a few tough months for Mom. Aunt Blanche didn't help any – she could be mean. (There is always a mean Aunt in every tale.) She told Mom that her husband was never coming back, but Her Majesty got Her provinces straightened out and Dad and the letters arrived in New Brunswick at the same time. Blanche must have been terribly disappointed.

So, before I arrived on the scene the family had done a lot of moving around. The thing was, it was not always by choice. Bad luck and circumstances had been the reasons behind their semi-gypsy address changes, but perhaps there was also a genetic compulsion to travel. I would work on this DNA until I honed it into an art, graduating in 1960 and continuing on to gain my Fiddle-Foot Master's and PhD before I was done.

Chapter II
Fiddle Foot Training – 1st Semester

I was born in Bath, N.B. in 1939 and in September Canada was back at war. The general family consensus was that I started it. I always treated that as the joke it was meant to be, but I wonder, was I already a handful before I could walk?

Dad re-enlisted as a Captain in the Home Guard making him a participant of both wars – a fairly exclusive club. He spent the war as Adjutant Commanding Officer at various POW camps and as 2 IC (second in command) he ran the camps. (No self-respecting C.O hung around a camp in the Canadian wilderness when he could be dancing, playing bridge and drinking hot toddies at H.Q.). One of Dads more notable postings was at Camp Neys on the north shore of Lake Superior. I still have a model of a four-master sailing ship made by a P.O.W. from materials at hand – a real work of art.

With Dad in the army and Mom nursing we had lots of good food on the table, decent clothes and decent housing. We would live in 3 different rented homes between 1940 and 1946.

Sidebar: These were pre-Medicare times, but when mom worked at a hospital – any hospital – all family operations were free. Therefore, we all lost some body parts at a relatively young age as tonsils and appendixes were yanked whether they were troublesome or not. Sort of a pre-emptive surgical strike, you might call it.

Our second rental, the first I can recall, was a restaurant with living quarters on the main highway leading up-river from Bath. It sat on the north side of the Monquart River and we only had to cross a short bridge to be in Bath itself.

The converted store where we lived and the bridge over the Monquart River. **(pic by Nancy McQuade)**

We moved into this place in 1942. I would have been three by then, and that's about as far back as my memory goes.

The restaurant was closed, no doubt because with the war effort siphoning off young people, there was a labour shortage. There was no electricity as it had been shut off to conserve power so we made do with coal oil lamps and Mom had a wood burning range and an ice box in the kitchen. We had lots of space to ramble around in and the restaurant area served as a living/dining room, although with heavy blackout curtains over the large windows it was always gloomy in the daytime.

Me and big sister Beth, two-up on the family double-bar CCM bike.

It was my next older brother Rod's job to keep the ice box working in the warm weather. The Monquart joined the broad Saint John just across the highway and an ice house sat on the banks of the Saint John a block or two down-river. In the wintertime we could watch men putting up ice.

So Rod would take a little wagon down to the ice house and buy a block of ice. If he was lucky, someone would put it in the wagon. If not, then he would have to wrestle with it himself – the block was heavy, and he was only ten years old. Back at the house there would be an older sister or Mom to help load the block into the top of the ice box.

Rod says that his reward was much appreciated – a hunk of ice to chew on a hot summer's day.

There was always activity on the big river. In the summertime men in hip waders stood in the shallows and casted far out for salmon – big ones. I don't recall log drives, although Rod has told me there was a spring drive. There would have been no birch timbers going down-river now. Ships returning to England would have young men as ballast and too many would never make the round trip.

Above town we could see the road towards Glassville clinging to the side of a high hill. This was the road our Grandpa Jones and the boys had used to bring the birch timbers down thirty years ago and it was still as long and steep as it had been then. We would often see cars going up and down that road.

One summer's day I watched a large truck climbing the hill very slowly – obviously loaded. My brother Rod told me it was so-and-so's White – but the truck was red! It took more than a few months for Rod to convince me that while the truck was red, the manufacturer of the truck was White Motors.

Table Scraps
Curiosity I
It's tough for little kids to understand when older brothers talk in riddles. I overheard my oldest brother Walter tell a tale of an incident on the Muniac Bluffs (The main highway going up-river came down off the escarpment before reaching Perth/Andover and the road clung to the side of that hill – similar to the gravel road behind Bath.)

As the story was told, a trucker was climbing the Muniac Bluffs when his overheated engine caught fire. He pulled out the hand throttle, climbed out onto the fender, opened the butterfly hood and beat the fire out with his hat.

Now you might think that was a pretty-much straightforward story, but not for me it wasn't. I had only recently caught the concept of the red/White deal and now I was sure the truck itself was a Muniac Bluff. Again, it took some time for Walter to unravel my noggin. I guess I was already the last longhorn in the stampede, but I had learned one sure thing – life is rarely straightforward.

Curiosity II
On our side of the Monquart bridge the highway shoulder slope near the bridge abutment was mostly small boulders – different sizes, different rock types of many different colours – very interesting to a four-year-old kid.

One nice summer day I was on the rock slope prospecting for gold. I was peering very closely at a good-looking rock when a little green grass snake, about pencil-size, popped out, stuck his tongue out at me and I went ass-over-teakettle down almost to the river.

From that day forward, although I still liked rocks, I did not like snakes.

Curiosity III

Just across the highway bridge stood a general store owned by the DeMerchants. (What a salubrious name – DeMerchants' Store.) From time-to-time I would accompany a brother or sister to pick up some groceries. It was my job to pull the wagon, saving my siblings' dignity, but I didn't care – there was always something to see and the chance of a penny candy. A mighty slim chance, actually – extra pennies were scarce and the counter clerk, although kind, would only offer a treat if it was a cash sale.

There was no pre-packaged shrink-wrapped meat and such in those days. These items were wrapped in brown butcher paper and tied with a string. High on the topmost shelf behind the counter sat a large cone of white string that passed through little brass eye-screws overhead and hung down beside the roller-stand that held the paper. In the blink of a kid's eye the clerk tore off a hunk of paper, wrapped the meat, grabbed the string, and wrapped it around the parcel, forth and back and side to side. He then wrapped the string once around a finger and snapped it off. This was all done in a carnival-like showmanship manner and the string snapping got an extra flourish – it was pretty to watch.

At home the string and paper was saved – nothing was wasted in our house. So one day I filched a piece of string and tried the snapping deal. I just about lost a four-year-old finger and crossed "store clerk" off my future employment list.

Curiosity IV: a Sticky Situation.

In our kitchen, of course, was the wood-burning kitchen range. On this stove sat (always) three irons. Some of you will remember these. They were shaped like little boats and had one wooden-gripped handle which could be attached to each iron as needed. On ironing day the boats were moved to a hot spot and were rotated as the ironing went on. What really interested me was how the boat switch was done. There was a thumb button on the handle – press it and it released a little finger locked to the heavy iron – move to a hot iron, release the button, and like magic, the handle once again was attached. I wanted to try that but I was NOT allowed.

Now the plot thickens – figuratively and literally.

Blackstrap molasses was made locally by sugar beet farmers and was used for baking. It was too bitter for table use but every self-respecting cook had a jug of blackstrap in the kitchen. Every self-respecting grocery store sold blackstrap in old-time earthenware jugs with a carved wooden stopper. The blackstrap molasses was thick stuff. The ten-dollar phrase is "high-viscosity!"

Our mom's jug always sat on the floor beside the stove. I walked into the kitchen, I was alone and I took the opportunity to test the heavy little iron boat/handle deal.

I dropped the doggone thing right on the jug and smashed it to Billy-be-damned! It was a new jug and I watched mesmerized as a full gallon of that gunk spread over the kitchen floor.

I ran, but I couldn't hide, nor did I stand up very well under cross-examination. I got no supper that night, nor did anyone else – the clean-up was long and laborious.

And I was pretty unpopular for a few days.

The Merry-Go-Round

When I was four, Sister Beth took me to the Bath Town Fair and with the war on It was actually more of a town picnic, but it sure looked like fun.

There was a merry-go-round! It actually did go around but would prove to be not-so-merry. Beth bought me a ride (10 cents?) and I climbed aboard.

It was quite a contraption. A sturdy metal centre-pole had a steel plate at the top. Chains led from this plate down to a large outer steel ring and suspended from this ring on smaller chains were seats which looked like mini park benches with room for one kid.

The motor was a pony led by a teen-ager in a small circle inside the ring of seats. A single-tree attached to an arm attached to the centre pole turned the whole deal and when the pony got up to full plod, centrifugal force caused the seats to swing out and up. I watched a load of happy kids and when the pony was brought to a halt and the swinging seats settled down it was time for a shift change. Beth helped me up, the other chairs were loaded and off we went.

This was great! There was no safety bar, but the seat was well-balanced and I leaned back holding the support chains beside me. We picked up speed and now I was flying high in the air with the wind whipping my hair. We all probably looked like puppy-dogs with our heads sticking out of car windows, and at peak speed and peak fun, the main centre plate lost a bolt and slid down the pole!

We all crashed and burned – and what a schmozzle! Parents, sisters and brothers waded into the wreckage to search for survivors. Beth reached me in an instant – it was easy to find me – I was squealing like a stuck pig! I had landed face first, conking my noggin on either a rock or hard ground and Beth checked for broken bones before picking me up. She cuddled me and said stuff like, "There, there, you're all right," until I settled down to sniffs and sobs. Then she sat me down and said, "Why, you've got a lump on your forehead the size of a walnut!"

That set off another crying jag. Sure – my head hurt a little, but mainly I wanted sympathy – and it worked. She took me to the fish pond and laid down another dime.

(Incidentally – one poor unfortunate broke an arm in the crash.)

The fish-pond was a perfect headache reliever. I was given a Huckleberry Finn fishing pole with a six-foot line and a hook on the end. The "pond" was about eight feet square with six-foot walls covered with decorated cardboard. We fishin' magicians could not see how deep the water was nor could we see how many "fish" it held. To "cast" you merely had to stick your pole over the top and Beth had to help me. A few short minutes and my rod bobbed, which I ignored. Then it bobbed again, and Beth had to tell me I had a fish – reel it in! Was I dumb or what? I jerked my line out and right there before my eyes was a fish! Not a real fish of course – it was something-or-other either fuzzy or furry and was attached to a piece of cardboard which had been hooked by me on my very first cast! I wanted to keep fishing but Beth was out of dimes so we shared a five-cent pop and walked home.

All-in-all it was a pretty good day. I could brag on my first fish and I had a forehead walnut to show everyone. It didn't hurt anymore but I didn't tell anyone that because by pretending I was tough as nails I could get more sympathy hugs.

I actually believed I had snagged that prize all by my own self. In my four-year-old mind I could see through the cardboard walls and inside the prizes were two or three feet deep. That was why I wished Beth had more dimes, because I knew there was a big one waiting in there for me. And then Rod brought me crashing down to earth – not a painful merry-go-round crash – it was a reality check crash.

He told me there was a local teenager in the pond putting prizes on the hooks and that destroyed my mind picture. Then he told me that the kid had to be quick and careful. The fish hooks were real and he had been snagged more than once. Now I felt a bit better, because I could picture this kid ducking hooks while doing his job. Thank you, Rod.

Rod was never very much for fairy tales. In his world logic ruled and everything could be explained by logic backed up by demonstration. I guess he knew he had punched a hole in my imagination balloon, so a week or so after the fish pond deal he helped me catch my first actual fish.

There was a fair-sized hill behind our rental. On the top a small brook meandered between tall trees before tumbling down to the Monquart River. This was our target today. Before we left Rod put our own Mark Twain fishing gear together.

He cut a pole, borrowed some string from Mom's stash and with a small purloined safety pin we were good to go. It was all very Huck Finn and Tom Sawer-ish and when I later read those books I realized where Rod got the idea, but at that time I was in awe of his inventiveness.

We followed the brook up to where it moseyed slowly through the heavier timber and found a fishin' hole – the real deal. The brook ran slower here and at the foot of some shale-rock mini-rapids was a quiet pool – maybe eight or ten feet wide and three or four feet deep. On the upstream side a little bridge had been built using two stringers and decked with smaller poles. It was only strong enough to carry a walking man but it was a perfect place to dip our line.

We sat on the bridge for a while with our legs hanging over and watched little brook trout swimming in the crystal-clear, cool water below us. In the shady parts they looked brownish but when they swam near the surface where sunbeams found their way through the trees they became silvery rainbows and we watched them chase water striders and swimming bugs. This was so idyllic, so quiet and peaceful – but it was time to get fishin'.

Rod found an angle worm and stuck the poor thing on the safety-pin. I dropped it into the pond and within minutes pulled out an eight-inch brook trout – I guess those fish were hungry.

We unhooked the poor thing and laid it on the bridge. I felt a little sad as I watched it die, and although it still had a slight silvery hue, the rainbow colours faded to brown. Rod, as usual, told me about the rule of survival – eat and be eaten.

We had half a worm left and now the other fish were wary, staying well away from the bridge. After a while we gave up and went home with the fish in my pocket.

Rod gutted and scaled the fish telling me that some fish were filleted, but brook trout were easily scaled. Then it was put into the ice box to wait for <u>my</u> supper. When Mom came home from work she fried it in butter for me and it was tasty!

About two weeks later I went back to the pool on my own. I didn't take any fishin' gear – I just wanted to watch the little trout enjoying fish life, but when I got to the pond I saw that disaster had struck. I guess the land owner was cutting firewood and had crossed the bridge with a caterpillar tractor. The whole thing was trashed and mashed into the brook and the pond was empty – not one fish had survived.

I walked home very dejected, with a hard lesson learned. Nothing ever stays bright and shiny – not in my world.

Our last move in New Brunswick was to Beechwood, six miles up-river and up-highway. There was no town and no post office, just a one-room school not far from our house. This move was in the summer of '45 and little brother Frank had recently joined the family in '44. Shortly after our move sister Judy was adopted, completing our family of seven kids. Dang it all – two rugrats to divert attention from the main star – me. This really put my nose out of joint and I had a lot of time to myself now. Older brother Walter was away at Dalhousie University in Halifax, eldest sister Jacquie was working somewhere, second-eldest sister Beth was looking after Judy and Frank, and Rod – five years older than me, thought I was too boring to play with. Likewise, I found the two youngest were also boring and with no other kids nearby I did my own thing – mainly trying to stay off everyone's hit list.

Curiosity V

I think it was our first winter at Beechwood. I would have been five – coming on six and Rod was building a model airplane. He had gotten the kit for Christmas and it looked like quite a project. It would have a three-foot wingspan, a fuselage in proper length ratio and was either a Spitfire or Hawker Hurricane. It would be powered by a long rubber band (supplied) connected to the propeller and anchored way back in the tail section. He worked on it all winter and I watched (don't touch) impatiently. It's maiden flight was scheduled for the first snow-free spring day. I could hardly wait (and I didn't.)

I suppose I could blame it all on Rod, He was always willing to answer my questions, the main one being "What makes it fly?"

He had finished the wings. They sat, ready to be attached, covered with fabric and sporting the RAF circles. The fuselage ribs and bulkheads were in place, as was the rubber band – only the fabric was left to be done.

And then Rod made a fatal error – he showed me how the prop worked. He twisted the band a little bit to show how the spinning prop grabbed the air, then went off to do some chores, making sure I went along.

I could not get this prop thing off my mind so the first chance I got I sneaked back up to his bedroom to try it myself. At first I just gave it a few twists, but then I got braver. I wound and I wound until the tail decided to meet the prop - and what a mess! Balsa-wood ribs and balsa-wood bulkheads were all bulged and broken and the fuselage was half the length it should have been!

I studied the destruction, knew I could never fix it myself so I found a place to hole up and cry. It was a practise cry – I knew a licking was coming.

I never did get that licking but it would have been the preferred option. I had to live with Rod's silent, devastated look of dejection, and my two older sisters' own disappointment in evil me. A licking would have been kinder.

Rod tried to fix it and a maiden flight was attempted. I watched with my fingers crossed but with structural rigidity lacking the wings folded up and the sleek fighter crashed and burned. Rod has told me he has forgotten the whole deal. Not I, nor will I ever forget.

Curiosity VI: Dan'l Boone.

There was a big old rifle in the attic. We were storing it for cousin Charlie and one day Rod brought it down for show and tell. It was not a flintlock musket but it must have been manufactured shortly after the flintlock. It was a muzzle loader and was a percussion cap-and-ball arrangement. The hammer on it was huge.

Rod told me it had a Damascus Twist barrel. Straps of heated steel were wound around a rod and when cooled the rod was pulled, leaving a smooth inner barrel. If you looked down the barrel you could see that the twists left "rifling" which, as Rod explained, helped the ball spin, which made the rifle more accurate. I was allowed to play with it, which I did from time-to-time. The older ones thought I couldn't hurt anyone or myself – could I? (Yes, I could.)

I'd hunt pretend bears in a little apple orchard beside the house. It was not an easy hunt – the gun was longer than me and almost as heavy. The only way I could line up on a bear was to lie on the ground with the barrel resting on a rock. Along came a bear. I dropped down behind my favourite rock, laid my cheek just behind the rear sight – I wasn't going to miss this one.

I pulled the trigger, and that huge hammer caught me under the chin. It drew blood and did it ever hurt – and the rifle went back into storage.

Sidebar: Years later I told that story to cousin Charlie and he remembered the rifle – he also had a story to tell.

Charlie had decided that he was going to test-fire the beast. He had no lead ball, but he dug up a steel ball-bearing that fit the barrel. Now he needed powder, which was donated by some shotgun shells. One shell supplied the percussion cap. I think Charlie said he used the powder from four shells before he paused to ponder. He knew the Damascus Twist had an evil reputation – there was always the chance of the barrel opening up near one's face, so he cut back on the powder load more than once until he felt things were copacetic. He dropped in the ball, (no wads – who needed wadding?) lined up on a tree a hundred yards away and pulled the trigger.

Anti-climax city! There was no loud bang – just sort of a subdued "blank" and the ball-bearing dribbled out of the barrel and fell in the grass.

And Charlie and I both laughed like crazy when the story was done.

Sling Shot

I was now six years old and a gentleman who could be trusted around glass window panes, so Rod made me my first slingshot. Nowadays, commercially produced slingshots are readily available and may even have laser-beam sight options, but in 1945 you made them from scratch, and it was not a simple task. As always – Rod made sure that I followed and understood each step during construction.

The basic weapon frame came from one of the old apple trees behind the house. Apple wood is strong and flexible and he chose a perfectly-shaped "Y". The handle was the right size for my small hand to grip and the prongs were just large enough to allow for small notches to hold the rubber strips. An old leather mitt was carefully cut to make the stone pocket with two slits cut on each side to pass the rubber strips through. This was the easy stuff.

Now comes the hardest part, and the most important part – the rubber strips.

Post-1939, natural rubber inner tubes became scarce. Rubber was needed for the war effort and when the last in-stock tubes had been sold synthetic rubber replaced them. This early bogus rubber stuff was rubber in name only. It had no stretch-ability and when tubes were installed any crease became a crack and every crack became a leak. Patching was ineffectual, so savvy owners of rubber-tired transportation retrieved their cast-off inner tubes, even scouring the dumps for the real thing. Leaky tubes were patched and repatched so when they were finally tossed away they had patches on patches.

Rod found one, and by carefully selecting unpatched areas, cut two – only two strips of slingshot rubber from one car inner tube.

Now I had a dandy weapon, needed ammo, and once again I was instructed by the future Civil Engineer. Proper-sized ball bearings were the best and we scrounged up a half-dozen. Those six were used for test firing by the manufacturing division and were lost.

That left stone ammo, and another class exercise ensued. Stones must be the right size and shape. Stones that had angles could cause deflection – Jeez – sling-shooting was high-tech stuff. So I collected proper stones and was turned loose with my weapon and with target restrictions. Windows and bottles were out. Little brother and sister were a protected species, as were the cat and dog. Birds smaller than crows joined the no-shoot list and crows were too smart, so I used tin cans. Tin cans on a boulder or fence post became enemy helmets. – fun for a while. Then I found the half-worm-in-the-apple and I had all the targets a boy could want. I picked off apples until the sling-shot was put away for the winter.

Sidebar: The little apple orchard in Beechwood was old and diseased. If you picked an apple before it ripened it would be scab-free, but I learned about the "green-apple quickstep, which needs no further explanation.

A little later on I was brought up to speed on the "half a worm left in the apple after you chewed your first bite" so I quit eating apples.

We moved west the following year, the slingshot was left behind, so in Manitoba I tried to make my own weapon, but willow crooks were not as strong as apple wood, natural rubber was still hard to find, and with no gravel around, I had to use old machinery nuts for ammo. By the time we moved back to sling-shot country I had lost interest in that kid stuff.

Table Scrap

Around 2002 – 03 my wife and I were tooling around southern Sask and we checked out the town of Avonlea. We went into the Co-Op Hardware, and tight there on a shelf was a dandy sling-shot with a "pistol grip," tubular rubber firing mechanism, and for mere pennies (with inflation factored in) you also got a tin of 200 ball bearings.

My wife had bragged on being a sling-shot sharp-shooter in her younger days So I bought it for her – an anniversary present.

(Give me a break – what else can you give a woman who has everything?)

Those early times in New Brunswick were learning years for me. For instance – I could read at four years old – it was a sink-or-swim deal.

We read a lot in our family – no matter how poor we were there was always something to read. We had a well-stocked bookcase and Dad still kept up our subscriptions – the Family Herald and Weekly Star, and the Saturday Evening Post among others. There were always newspapers – not necessarily every daily paper, but we always got the Saturday edition and on Saturdays there were coloured funnies!

In Bath every Saturday I'd bug Rod to read me the funnies and he hated doing so. Finally he said, "If you want to read the funnies you'll have to learn to do it yourself," so I did.

(Later, my rugrat sister and brother bugged ME and I found out why Rod hated it. There is nothing better to take the fun out of funnies than having to read them to illiterate little kids. Judy and Frank got their own ultimatums.)

In 1945 I started grade one at the little school down the road. I thought it would be easy, since my oldest sister Jacquie was the teacher – how wrong was I! The problem was, that she would praise me in front of the whole school, and I was soon called a "teacher's pet." It was a sure death sentence at the hands of the big kids at recess.

So I released my evil alter-ego, and now I was in danger of getting the strap. Wow – on one hand the strap – on the other hand a recess licking – the scales of justice are not easily balanced.

Jackie would not strap me but the hurt look in her eyes when I was bad was punishment enough. Finally, one evening she cried at the supper table and now I had the whole fam-damily to deal with. So we hammered out a détente – I would be good as long as she quit drawing attention to me. What I did not, and would not tell my family was that I knew I could talk tough at recess as long as there were no tattle-tale girls around and I successfully walked that tight-rope for the rest of the year.

I have other fond memories of New Brunswick – picking fiddleheads in early spring, wild strawberries and raspberries in summer and collecting beech nuts in the fall. The winters were soft. and with lots of snow and hills, my Flexible Flyer was seldom idle. The war was always front and center on the radio and in the papers. We had blackout drills – subs had been spotted in the harbour at Moncton. Looking back, I know it was small potatoes compared to what kids were going through overseas, but it was scary stuff for sure, because I had an older brother to prime my over-active imagination.

Dad mustered out in 1946 and a move was in the air. This was going to be a big one – back west again! The news slowly filtered down to the ankle-biters. Our tender psyches must be dealt with gently until we accepted the fact. Well not me, by golly! I needed no touchy-feely reassurance – I was ready to roll. This was no mere shuffle to a different house and I was impatient.

Our mother was not averse to the move either. In 1946 I was unaware of the line drawn in the Saskatchewan dust in 1938, but years later I was brought up to speed thanks to my older siblings.

The thing was, when mom returned to N.B. From Sask. she ran into a modified "You're not for here." attitude. Her family was no problem – they were always generally supportive – it was the society ladies in Bath that chose to cut her off. As an R.N. Mom should have been accepted as a professional lady in her own right, but the Christian Women's' Guild turned their noses sky high.

When my dad re-upped and was promoted to Captain, the local paper printed the news and Mom became a star! The Guild invited her to a tea.

So she went to that tea, dressed to the nines, including her Sunday-go-to-church hat, and the ladies were so glad to meet her! The president of the Guild came up to her and said, "You never told me your husband was an officer."

"You never asked," she replied. Then Mom told them what she thought of them and left, never to return.

It was a Jeannie C. Riley "Harper Valley P.T.A." moment and we were sure proud of our mom when we heard the story.

Table scrap: "You're Not From Here."

Saskatchewan is a fine province, rich in agriculture and natural resources. People are outwardly friendly but Saskatchewan is definitely a "You're Not From Here" place.

In the '70s I lived in Vermilion Bay, Ontario and was working as a compressor station operator for Trans-Canada Pipelines. They kicked off construction on the second line and Majestic-Wiley came to town. A local lad caught on as machine operator – he was a good man and the spread boss liked him. When Majestic moved to Saskatchewan they took Dennis along.

Less than a week later I ran into him in town – didn't he like the prairies? Dennis told me that he never had a chance to investigate. The day after he arrived he boarded the morning bus to the spread and someone asked him where he was from. They seemed to be an easy-going group of guys and he told them he was from Ontario.

The next day Dennis was not allowed on the bus.

Table Scrap: You're Not From Here Part II

Fast-forward to 2004: My wife and I had a leased restaurant in a small town an hour south of Regina, Sask. We did an all you can eat hot table brunch on Sunday, and most Sundays covered the week's overhead. As usual we were hands-on and I generally got in at 4 AM to fire up.

One Sunday had been exceptional and by 7 pm I was tuckered but pleased. We closed at eight and our waitress had already gone home. The half-deaf worst waiter ever only had two people to look after – an older couple I had never seen before, and the lady asked where I was from.

"I am from here," I said.

"No you're not," said she, and just like that I got into an argument I should have known I couldn't win.

Me: "I live here. I'm from here."

She: "You're not from here. I'm from here."

Me: "I live here. I own a house here. I pay taxes here. I employ people here. I'm from here."

She: "No you're not. I'm from here." (Broken record)

Me: "If you're from here, why haven't I seen you around in the four years I have been here?"

She: "I got married 50 years ago and moved to Regina. I'm in town visiting family."

I made one more effort. "So you live in Regina, own a house there, pay taxes there, raised a family there, are now retired there, and come down here once a year. Yet you are from here and I'm not."

She leaned back in her chair and oh so smugly delivered the final thrust. "Now you've got it."

I walked into the kitchen and told my wife the front end was now hers. I walked out the back door, sat down and ate worms until my wife locked up.

That pretty much did it for me. Six months later we closed the joint and moved back to Ontario.

I must add this foot note to the story: I've lived in the Rainy River Valley off and on since 1950, yet I have never heard the dreaded "you're not from here". Good for you, Rainy River Valley!

Dad had a plan and had been working on it before he was discharged from the army. His last POW camp was in Ontario and he bought an almost new 2-ton truck and a decommissioned railway water tower. These purchases may seem to be unconnected, but read on my friends, I will elucidate.

The truck was a 1945 Ford. On the assembly line it had been slated for duty in North Africa and therefore the engine was built to handle dust and blowing sand. It was called a "desert motor" and I think the modifications were a larger oil-bath air cleaner and an oil filter (oil filters were an option at that time). Whatever the mods were, they would prove to be effective because from 1946 to 1950 that Ford seldom ran on pavement. It was a huge truck as far as I was concerned. It had 750 X20 tires! (Nowadays any pickup worth its salt has 20 inch wheels).

Not ours: However, the fender dent is accurate. Maybe it was a factory option.

Hostilities ended before the Ford hit the assembly line paint shop, so it was painted green instead of desert brown. A local trucker bought it - his fleet needed upgrading and when civilian vehicles became easier to buy, he further upgraded to a 3 ton. Thus dad was able to buy the low-miler 2 ton.

Sidebar: It was a seller's market in post-war North America. Civilian vehicles in Canada had been unavailable since 1939, and now inflation hit hard. A new Mercury four door cost $850 in '39 and by 1950 a 4 door Merc would set you back more than 22 hundred bucks, so Dad probably paid a good dollar for the 2 ton.

Now we get to the water tank. The tank itself was all hardwood - oak, ash and/or hickory. The tank uprights were 14 footers, 3 inches thick. These were re-sawn to 1 1/4" and were used for the side boards and the headboard. Some were sawn into 3 1/2 x 1 1/2 tongue and groove for the deck. The oak (or hickory) became stakes and the side boards were bolted to them. Three-inch angle iron surrounded the deck with stake pockets therein and a conventional double tailgate completed the deal. It was a work of art – heavy, solid and convertible. The sideboards could be removed and the deck would haul pulp, hay bales and what have you for years to come. During the water tank's railway life, it had sat on a base 20 feet from the ground, supported by B.C. Fir timbers. These were used as barter for the box construction so Dad's actual cash layout was minimal.

The two-ton doubled our family fleet. Walter had turned sixteen last year and Dad had purchased a 1937 Plymouth. So far, the car had been used only to take Mom for groceries or family visits. But it had not been bought solely for that purpose. We now had enough rubber on the road to enter the next family chapter.

Chapter III
Fiddle Foot Training - Second Semester

We were heading west – destination, Manitoba!

I had no worries about the move. We had already moved three times since I was born, so another one was no big deal. I had completed grade one by then, but had no close friends to leave behind. I did have a bit of separation anxiety – two things were worrisome. One was that our oldest sister was staying behind and I would not see her again until she was married on our front yard in Manitoba. The other loss was our cabinet radio – a big old Philco with FM and short wave. It also had one other switch – radio or TV! We had heard about television but had never seen one. Sometimes, when no one else was around, I would select the TV mode and imagine what it would look like. I was told that there was no room on the truck for the Philco. What I didn't know was that we would be electricity-less for the next seven years.

We pulled out in early August – all eight of us – three in the two-ton and five in our 1937 Plymouth four-door sedan.

Our oldest brother, Walter, drove the Ford – he was eighteen and all growed up. I rode in the middle, as I was small enough that my legs didn't interfere with the gear shifter. My second oldest sister, Beth, usually rode shotgun and I liked the arrangement at first – they were both my heroes – but the roads were not exactly smooth in the '40s, and I bounced a lot. It didn't take too many days before I wished I could trade with my sucky little brother and sister and get to ride in the relative comfort of the Plymouth.

We led the way in the two-ton. It had no two-speed rear axle and although the speedometer was marked to 60mph it was an unattainable speed – 30 mph was usually the max. It frustrated the dickens out of me. I would watch that needle hovering half-way to 60 and imagine zooming along at that speed.

(When we hit the Shield west of the Lakehead there were some long hills. The Captain behind us had given Walter his orders – no freewheeling allowed. But on one long hill, with the Plymouth far behind (or ahead) I talked my brother into trying Georgia Overdrive. We hit 45! Wow!)

We entered Quebec and passed neat farms and neat towns with large churches. Gas stops were restroom stops, and there must have been quite a lineup at the facilities. Towards suppertime we would find a place to camp. Four long 2x4s and a spare large canvas tarpaulin were removed from the tarped load. The 2x4s were leaned against the truck box and covered with the tarp, making a lean-to for us to sleep under. A two-burner Coleman was our cook stove. (Did we have two Colemans?) It must have been quite a chore to boil potatoes, a vegetable and fry meat for eight. At bedtime we all snuggled in like pigs in a blanket. The two littlest ones usually slept in the Plymouth with an adult for company.

It often rained at night and I liked the sound of raindrops on our roof. An added bonus was the aromatic smell of damp canvas. I'll remember that always, and although I did not know it in 1946, I was already being trained for bush life. Fourteen years later I would be sleeping in a tent every night – and I liked sleeping in a tent long before I actually lived in one.

Sidebar: Years later Walter showed Frank the stoves – there were two – sold by Sears, and a little larger than your regular Coleman model, with an added feature – a filter for lead, meaning they could burn automotive fuel in a pinch.

Farm after farm passed by slowly, soon becoming same-old, same-old. The bridge across the St Lawrence River at Quebec City was interesting and in Ottawa we got to climb the Peace Tower at the Parliament Buildings – little ant people and toy cars far below. I felt drawn over the edge and quickly stepped back.

North of Renfrew along the Ottawa River things got more interesting. There were more pines here, a few outcrops and the farms were further apart. North of North Bay we went through some rugged Canadian Shield, and after Cobalt we returned to farmland in the Earlton clay belt. The terrain flattened out with conifers interspersed with farms, and west of Cochrane to Kapuskasing it was much the same. After Kap, the roadside was just miles of lowland spruce and muskeg. On the west outskirts of Hearst, we passed that famous sign – "WARNING – NO GAS FOR 120 MILES."

(That sign would remain until the '70s, when a tourist camp finally put in gas pumps halfway between Hearst and Longlac.)

Table Scrap

The two-ton was never meant to be a highway hauler so it had no saddle tanks and the gas tank, which was, believe it or not, under the non-adjustable seat, was no larger than the tank in any Ford pickup of the time. The 120-mile stretch from Hearst to Longlac was the longest gasless piece of road, but gas stations were (and still are) few and far between, thus two surplus army gas cans were part of the load.

One day they must have pushed it hard, intending to extend that day's itinerary. The Captain had gone ahead to scout out an overnight camping spot. On a long hill the Ford ran out of gas and the jerry cans had already been used up.

Walter knew the last station was miles behind, and he also knew it might be nightfall before the Plymouth came back, so he decided to take a can and walk ahead. Maybe he could catch a ride. Fat chance – it was a lonesome road in 1946.

He walked over the hill and in the other side was a welcome sign – White Rose!!

Table Scrap: The Micro-connections of Life

I was seven years old when we crossed the St Lawrence at Quebec City and as it was the biggest and most remarkable landmark of the trip so far, I paid attention. For one thing – it carried trains and vehicles on two decks. I remember that we were on top and a freight train rumbled beneath us. I also recall that it was of "meccano- like" construction, with enough girders, braces and connections to keep a kid working for a year, and to keep Meccano in business and numbers forever.

(Meccano sets were bought in progressive numbers, "0" being the starter kit. One Christmas I got a "0" but by the time another Christmas rolled around and I opened a "1" I had lost most of the "0" parts.)

Now we must hit the fast-forward button to 1970. It is January, I am at the OCA base in Red Lake mobilizing for our return to Black Beaver, and I See OCA's storage area is full of pieces of aluminium and such that looked like parts of a trashed aircraft.

Dave Harvey, the base manager, tells me the parts are from a Fairchild Super-71 that had hit a "dead-head" (submerged log) at Confederation Lake years ago and had sunk in the lake. At the time it had been pulled from the water to recover the gold bars it had been carrying and the wreckage had been left on shore. Now, with the road into the new Selco Mine, Dave and some others had been hauling the parts out with rebuilding

the Fairchild in mind. It was (and is) a rare example.

They soon found the project beyond their means and donated the stuff to the Western Aviation Museum in Winnipeg, where it is now on display, looking

factory-fresh.

Anyway – Dave told me that the reason the pilot hit the dead-head was because he could see dick-all ahead of the plane due to the cockpit being so far back in the fuselage. He also told me that one unfortunate had wiped out a Fairchild at Quebec City. The seaplane base on the St. Lawrence usually did their takeoff runs beneath the bridge, and the pilot had confused the bridge bars and girders with the braces he had to peer through under the Fairchild's "parasol" wing.

So – 24 years later I recalled my own bridge crossing and I understood that pilot's predicament.

Near Geraldton (gold country) outcrops began to appear again, and west of Nipigon there was lots of interesting stuff along the north shore of Lake Superior. Northwest of the Lakehead we got back into the Shield. This was my cup of tea for sure. The highway was now up and down, curving left and right, with muskeg and spruce on the low ground and ponds and blue-water lakes in between, morphing to jack pines and red pines on the ridges. Now there were rocks and rock cuts – everyone different and around every corner a new vista! These were rocks of every colour – grey granite, rusty iron formations and darkish-coloured greenstones. I didn't know jack about rocks then, but I sure liked this.

(My only previous experience with rocks was that if you found one the size of an oatmeal cookie, flat, with nice round edges – and if you had a good arm like the older boys, you could skip it up to lucky thirteen times on a still pond.)

West of Dryden we passed through Vermilion Bay and I'm not lyin', nor is it a faulty memory, but I liked that little town. I liked the way the houses nestled in amongst the jack pines on a sand flat high above Eagle Lake and I really liked the cozy look of a log cabin motor court. Vermilion Bay stayed in the back of my mind for a long time but it would be thirteen years before I would pass it again.

Ten miles east of Kenora we passed the junction with Hwy 71. That highway would play in our future, but on that day it was just another road.

Into Manitoba we drove, through West Hawk Lake and on to Rennie. Twenty miles west of there we left the Shield, the pines, the lakes and the rock cuts, and broke out into the prairies. It was not a swift break, but as each mile passed the spruce trees became smaller to be replaced by poplar bluffs and farm land, until it was all farm, all prairie, all the time – and not a hill to be seen to slide on. I knew then that me and my Flexible Flyer were in trouble.

In Winnipeg we stayed at a motor court within sight of the Manitoba Legislature while Dad and Walter looked for a suitable farm. The cabins sat beneath lush elm trees and we needed two to hold our herd. They were a comfortable break from the canvas covered lean-to and it was a very quiet oasis even though we were on a busy street. As dusk turned to dark we could hear the street cars passing almost silently with cheerful ding-dings as they warned stray jaywalkers. When they crossed the Osborne Street bridge there was a louder, hollow echoing rumble – a peaceful lullaby for sleepy little kids.

It must have been a welcome break for Mom – a chance to relax, re-group, launder three weeks worth of travelling clothes and to hose down three stinky kids, but cheaper temporary quarters were necessary so an empty farm-house was rented at Argyll, twenty miles north of the city.

At Argyle they unloaded the two-ton but only unpacked necessities like clothes and kitchen stuff. Boxes and bags were piled in one corner of the living room. One bag held trip-related items among which were six or seven rolls of Kodak film. These held the trip history, awaiting development.

.

Disaster!

It was an early September afternoon and I had a job – I was to watch Frank and keep him out of trouble. Judy was helping mom in the kitchen.

We played outside until it started to rain, then we retreated to the living room. Frank sat on the floor, pulled some articles out of a trip bag and seemed to be quite content studying on different items. I found something to read and was soon in my own little world – not a great multi-tasker.

Mom came into the room and screamed! My head popped out of my book and there sat Frank – happy as a pig in poo surrounded by coils of undeveloped and now exposed film. Every picture was ruined and the trip record was gone!

I'll bet you can guess who got the blame and the licking: and even now, 69 years later, it still pisses me off.

Frank on a rock near Kenora Ontario. This is one of two pics that survived because they were still in the camera.

(From Frank's Perspective)

I don't recall the unrolling - I have to take it on faith that I was there – but I sure recall the aftermath. For ever after I was reminded from time-to-time of the terrible feat I had performed in childish glee. Every sibling but Bob - who had the most right, since he got blamed, and Judy - who stuck up for me no matter what, would remind me from time-to-time. It was not really an accusation, just a forlorn memory of a valuable heritage record lost forever. I eventually learned not to defend myself – just let sleeping dogs lie.

The Argyle farm house back-story has always been a mystery to me. It was a solid building, and across the driveway was a good-sized farmyard with many trees to shelter the place from winter prairie winds. There was no machinery around – none at all, but there was a huge barn. It was T-shaped and the cross on the tee was, I'm sure, one hundred feet long. Cattle stanchions lined each side. There was room for at least fifty cows and it had good cement floors and huge haylofts. It had obviously been a dairy at one time but there were no pastures or fences. All the surrounding land was either stubble or in summer-fallow.

The whole place was now owned by a farmer a few miles away. I've always wondered – was there a war-related story here? Did the previous owner go overseas, never to return? Perhaps the family lost some sons and could not carry on. Whatever the reasons for the place being empty I really don't want to know. Some things are just too sad to dwell on.

Brother Walter told me many years ago that the farmer/owner wanted Dad to stay. He would back Dad the following spring to put in a half-section of flax, but Dad turned down the offer – he no longer trusted open prairie.

The guy sowed the flax himself. He took off nine bushels per acre and flax was seven dollars a bushel.

Rod and I started school in Argyle in September and Argyle was five or six miles away so we rode the school bus, if you could call it that.

It was actually a Ford Model A two-door and there were four other kids in it when we got picked up, so we were packed in pretty tight. The road was not gravelled until we got to town and when it rained the Model A was replaced by a horse pulling a van which sat on a rubber-tired farm wagon, in which we sat on benches facing each other. I was told that in winter-time the van body was shifted to a sleigh.

The school was a huge two-story deal with two large classrooms on each floor. Because it was grade one to twelve the lower level was for us little kids.

I have only two memories of that school. One was of the geography book we used. I can picture it still – a large gray-covered book with liberal use of maps and photographs, one of which showed "The largest out-door wood storage in the world." I didn't know in 1946 that 20 years later I would see that yard in International Falls, Minnesota every day as I worked at the paper mill in Fort Frances, Ontario.

The other memory is also about that same book and a mean old teacher. We were studying geography and I was talking instead of listening, so I got a geography lesson. She ambushed me from my rear quarter-flank, grabbed the book from my desk and fetched me a good, hard whack on my right ear. To this day I have a pin-hole in that ear-drum. I didn't complain about my sore ear at home that night. I didn't want to put my other ear in the line of fire.

In late September we moved to our new farm at Deerhorn, 75 miles north of Winnipeg on Hwy #6 and I was glad to leave that teacher behind.

Before we left Argyle Dad sold the Plymouth – a grave disappointment to me and for sure, to our Mom. The car, bought in 1945 for $400, was sold for $450 and part of the proceeds went into the final mod for the two-ton, thus the Plymouth served a dual purpose. It not only helped move our family westward – it also supplied the cash for that final, very important truck-box modification/invention.

(During the final two weeks before we moved I would occasionally catch a glimpse of the car with its new owner behind the wheel and I decided that I didn't like him very much. How about that? I had a hate on for a man I'd never met.)

Dad and Walter used part of the Plymouth cash to build, with the help of a local blacksmith/welder, an all-purpose grain auger. It had to be hand-fabricated because it was unique and unavailable at any machinery dealer anywhere. I won't go into any great detail, but when completed, that auger, powered by the two-ton's PTO could unload wheat and such into a granary or boxcar and also would be able to self-load the truck from any storage facility.

After all that good planning and good workmanship to build a dandy farm truck, Dad had to pick a farm where we were almost guaranteed to slowly starve to death. That truck saved our bacon, though, because starting that fall of '46 and each harvest season for the next four years Walter ran that 2-ton 24/7 hauling grain for 2 ½ to 3 cents a bushel. Some years it was the only difference between having a good Christmas or coal in our stockings

Dad picked the Interlake because it was parkland instead of bald prairie – he wanted some trees around to help hold the water table stable. The trees were disappointing to me, mostly poplars of all sizes. The only evergreens I saw were planted in the odd yard here and there. Our Christmas trees had to be imported from 25 or 30 miles north. The only rocks were stones in the plowed fields and they had to be picked – pure drudgery.

Table Scrap
A city family is out for a drive in the country. The little kid says, "Daddy – look at all those rock piles in that field!"
His dad replies, "That's a sign of a poor farmer. He's too lazy to spread his stones."

Same song 2nd verse: This was an ad in the Country Guide circa 1993. "For sale: 80 acres in Bruce County near Owen Sound, Ont. Fields contain enough rocks to build your own fences."

The house in Deerhorn was a big old square two-story job made of hand-squared poplar timbers, chinked with plaster outside and in, and the inside walls were covered with some sort of wallboard. There was no insulation, and over the years the inside chinking had come loose in spots making bulges in the wallboard. Brightly-coloured wallpaper with lots of flowers thereon disguised the imperfections and made the place cheerful and homey. The whole house was heated by a big 36-inch airtight stove in the dining room. There was one drawback – the outside walls were slowly sinking on their stone perimeter foundations. You couldn't play marbles in the house – they always rolled out towards the baseboards.

Don't get me wrong – the Manitoba Interlake had its own charm. We had soft summer evenings with whip-poor-wills calling at twilight. There were other birds I had never seen before – yellow-headed blackbirds, western bluebirds, and meadow larks

46

singing while perched on fence posts. (Where did meadow larks perch before settlers built fences?)

We still picked wild strawberries, saskatoons replaced blueberries, and wild hazelnuts were a tasty stand-in for beechnuts. Winters were colder and windy. There was probably as much snow as we had in New Brunswick but the flatland winds would not let the snow lie where it fell, continually moving it behind buildings and fence rows. We did get the occasional tail end of a Chinook – I liked Chinooks.

The road to our new farm was something else – mostly unimproved, you might say. One-half mile north of the little village of Deerhorn we turned left off #6 Hwy, crossed a slough, and virtually disappeared.

Sidebar: Terminology heads up: I'd never heard of a slough before. In New Brunswick we had swamps – on the prairies and in the Interlake we had sloughs. They were of various sizes and larger ones had open water in the centre. Smaller ones were covered with reeds and cattails. Ducks liked the open water and blackbirds liked to perch on the cattails. When we moved to Crozier (near Fort Frances,) we were now back in Ontario and I had to change my thinking back to "swamp." No one in Crozier knew what a slough was.

The first ¼-mile across the slough was built to grade (no gravel) and it ended at the tree (or bush) line. The next ¾ mile was a combination of short sections of grade (statute labour) and higher ground dirt sections with detours around wet spots. In dry weather one could drive straight through but after a rain a new track was made around the edges. If the weather was really wet or if the ruts got too deep another wider arc was used. If you could see these from overhead, they would look like knots on a board. Past Joe Rivet's corner the last half mile to the school was very much a turkey-track (if a turkey was dumb enough to use it.) On past the school, the last 1/8 mile to our house was again a statute labour grade.

(Statute labour was a way to work off your land taxes and thus was popular during the depression. There still may be statute labour bylaws in some municipalities, but I don't think it is done much anymore.)

So that was our road in the snow-free months. In the wintertime it changed character – generally to impassable. Then the two-ton would be left at the bush line and the horses and sleigh would take whoever was desperate enough to head townwards – usually to the little store in Deerhorn. If the truck was needed to go six miles north to Eriksdale or seven miles south to Lundar, the horses would be tied in the shelter of the bush and were given some hay to munch on.

Winter or summer, it was a Manitoba Interlake version of gridlock, often taking us longer to get to the highway than to actually drive anywhere.

We walked a ¼ mile to a one-room school which, for some reason, had been plunked not far from our house, and all the other students had longer distances to travel. Reg Henrotte walked through the bush from their farmstead one mile south – two Monkman kids likewise from one mile north and Harvey

Rowland also came through the bush west of our place. From Deerhorn and a couple of farms near the village came the Lowrey and Richaud kids. They had almost two miles to walk, but they had a road (?) to use.

Robert Houston, two years older than me, came from two miles to the northwest. He had to skirt a large slough before hitting a bush trail. He seldom had to walk and he never travelled alone – the teacher always boarded at Robert's house. Robert's uncle was the chairman of the school board so Robert and the teacher often had a horse and cart in the summer, trading in the cart for a toboggan in the wintertime.

We were never crowded in that school. There might be as few as nine students and never more than twelve. If one kid graduated from grade eight there might be two to replace him/her in September, and if two graduated only one would start the next year. One year a CNR section man moved to Deerhorn and they had two kids. They were gone the following year – no married section men were eager to move to Deerhorn.

Throughout my elementary education I went to one-room schools and I still think there was nothing wrong with them. When you got bored with stuff you were doing there was always something interesting going on in the higher grades. Any graduate of a one-room school has been in grade 8 eight times.

The original Deerhorn School, later replaced by a frame building.

The school was of normal frame construction and seemed to be well-built. It sat on a low stone foundation, had high multi-paned windows on the south side and even had a bit of a porch and a cloak room. (Hooks on the entry-way wall.) To the left of the door to the single classroom was a little stand with a pail and wash-basin. One communal tin cup hung on the wall above the pail. I cannot recall ever using the wash basin – there was no soap or towel provided.

The desks were doubles and were free-range – no screws held them to the floor. This would come in handy in the coming winter.

There was a large black-board on the south wall behind the teacher's desk. Higher up on that wall was a picture of King George VI. and we watched him every morning as we sang an off-key "Oh Canada". Did we ever sing "God Save the King?" It would have to be up to God – not one of us kids gave a rat's ass.

A smaller blackboard (actually a green board) hung on the north wall and I recall there was a picture of a sailing ship either to the right or the left. Why a sailing ship, I do not know – perhaps it was donated to cover a hole. This was the sum total of our wall decorations.

A big barrel stove sat in the middle. The stovepipes went up to a 90-degree elbow and headed south beneath the ceiling, to disappear into the exterior chimney on the outside south wall. This was typical of those times – no heat was wasted and we would need that heat on cold winter days.

The school yard was also typical. On the west side of the building was the long wood pile – seasoned three-foot poplar which was put up every spring by way of a wood-cutting bee. The poplar was donated by the King from a Crown ¼ section on the north side of the road. Just south of the wood pile was our sanitary (?) facilities – two separate one-holers. They were not unisex nor did they sport a helpful sign. It was like figuring out which were the boy or girl sardines – you had to watch which can they came from. I think they may have had an Eaton's catalog available, but mostly it was BYOTP.

The "campus" was on the east side – a sizeable hunk of grass surrounded by willows and scrub poplar. We had a baseball diamond of sorts. Home base was near the water pump – a handy water-cooler. There wasn't much of an outfield and a strong lad at the plate could hit a home run into the bush if we'd had a decent ball to hit. I think the school board got tired of buying a leather-stitched softball every year so an imitation rubber-clad ball was what we used. It soon had the structural integrity of a rotten orange, and a home-run swing might moosh it past second base. Home run swings were pretty tentative anyway, as our one split bat was held together with hockey tape and even the tape came from who-knows-where. The nearest open natural ice rinks were at Eriksdale or Lundar.

Most of us were poor folks. Some families were on welfare and others, like us, were subsistence farmers. You might envy a neighbour who had a better farm or a new tractor but no one had money to burn – nor did the school board.

Our teachers were of the "supply" variety – not yet fully trained or accredited. They were probably paid less than $30 a month and there was a new one every year. They were always young, always kind, and while the strap lurked, I never saw it come out of the dreaded drawer beside the teacher's right leg. We understood the threat and while we could be bad kids, we were too unsophisticated to be really bad. We were simple hicks and I know for a fact that sometimes we drove the teacher nutsy.

We made our own fun. Playing softball was boring, so every fall we had the "ANNUAL GROUND SQUIRREL HUNT." We called them ground squirrels but I think they were smallish prairie dogs – (My brother Frank now tells me they were Richardson's Ground Squirrels.) After the firewood was piled up in the spring the varmints would nest there and raise their families, so when we returned to school in September we had lots of cannon fodder waiting.

It always started out innocently enough – we might be playing tag, pom-pom-pullaway, or red rover, why don't you come over.   Sure enough a little ground squirrel would pop out of the wood pile and the hunt was on!

Everybody, boys and girls, big and little joined in. Sticks were found, the woodpile was shaken and poked and soon there was a squirrel for every kid. They headed for the bush, and so did we. Recess was over, and the teacher was ringing the handbell, but even if we heard it, we ignored the clangs.

Eventually we powered out and struggled back to our irate teacher for a lecture. The dressing-down never stuck. Next week the hunt might break out again and I guess the teachers learned to live with it. One thing for sure – after the hunt we were so tuckered out we would be no trouble for the rest of that day.

No ground squirrels were ever hurt – they were too fast, and we were too slow. However, one day when I was hot on the trail, my victim ran between Robert's legs and I busted my stick on his head. Lucky for me – I had grabbed a half-rotten dry poplar.

Robin Hood was also a favourite, usually initiated by me or my best friend, Reg. Robert always had a good jackknife in his pocket to cut proper willows for bows and smaller straight ones for arrows. We had a roll of binder twine at home, so I would borrow (?) some to string the bows.

We had rules – blunt arrows only and no head shots allowed. We had to bring the girls in for foot soldiers so boob shots were also out – girls had to take an arrow in the back.

What was unfair was that Reg or Robert always got to be Robin Hood. I had to be the Sherriff of Nottingham or King John and I never survived.

And of course, Robin Hood and his Merry Men lived in Sherwood Forest and once again, the bell was ignored. I sometimes felt a bit sorry for those poor teachers.

In the spring of '49 we actually had a shop project. It was our own idea and that year's teacher was left out of the loop.

The plan was hammered out at a recess board meeting. We would build a log house. Reg and I were ten and Robert would soon be in grade eight, so he was the Alpha Male and would be the go-to guy. Reg and I were sub-foremen, the girls and little kids were roped in and no threats were needed – everyone was gung-ho.

A site was picked in the heavier woods in the crown quarter across the road. We found a level dry spot with suitable poplars nearby and far enough in so that Teach couldn't hear us working. It would be pure pioneer stuff – no saws or nails, but we needed axes. Robert supplied one, and I another, and this required covert manoeuvres on my part. I don't know how Robert handled his, but I stashed mine in the brush beside our driveway one evening, carried it to school and stashed it again. Frank would start school next fall, so I only had to tell Judy to keep her mouth shut.

The work was done at recesses and lunch hours. Judy and I always had lunch at home and our mother didn't notice that we were eager to get back to school. The school-yard must have been unnaturally quiet but we were smart enough to answer the bell promptly now. If a day was windy we would post a half-pint halfway sentinel to warn us. There was a job for everyone.

We went to work. While some removed small trees and bushes at and around the building site, Robert dropped the first tree. We had planned this out at our frequent company strategy sessions. Trees near our cabin would be left for shade and when a tree was being dropped no one was allowed in that area. Robert made sure of that and a sub-foreman enforced the rule.

The building plan was also hashed out beforehand with all board members present. It was to have one door and one window and the walls would be six feet high. The logs would start with the largest at the bottom and slowly decrease in size as we went higher, so more than one tree was felled before construction started.

I don't recall what the original planned dimensions were but when it came time to cut off the first timber we realized eight feet would be the maximum size we could carry, so the blueprint was altered to six by eight feet (exterior.)

Someone's desk supplied a 12-inch ruler and the first logs, actually nine foot and seven foot (to allow for notching) were axed off – Reg and I helped Robert do this. The three of us also carried the base logs and they were darned heavy. Then Robert started notching while Reg and I continued cutting off more logs.

Now there wasn't much for the smaller kids to do. Robert, as always, had an answer. From branches we made three-foot carrying sticks with a little curve to cradle the logs. Then with four middle-sized kids – two on each stick, they could haul logs. Also, as each row was finished the wee ones collected moss and chinked the wall gaps – we were a well-oiled machine.

I had to borrow a hammer and someone filched some four-inch nails to provide temporary rigidity to the door and window openings. By working every recess and every noon-hour, when the weather permitted, we actually had the walls up to the top plate before school was out for the summer.

We knew we were in trouble now, anyway. We had no idea how to do rafters or roof, nor did we have boards or 2x6's to frame the door and window. Judy and I were the only ones who lived close enough to supply these materials, and spare lumber was as scarce as hens' teeth at our place.

But we were well satisfied and proud of our work and of course, we eventually got ratted out by some little kid. The teacher had to see this, so she got the red carpet tour with all of us in attendance.

When she saw those sharp axes and big stumps she almost freaked out, but we had not lost any arms, legs or fingers, so she soon settled down. She walked around checking workmanship, said we had done a good job, and ten or twelve chests puffed out. I'm sure she bragged on her baby pioneers when she returned home to Winnipeg.

And I'm here to tell ya – I have never had as much fun at any school than I had building that log house.

That fall, when we returned to school, the cabin was forgotten. We had a new teacher to train and new families of ground squirrels to terrify.

(No lies were fabricated in the telling of this story.)

I wonder now how any of those teachers could handle a complete year. They were paid little enough to teach us, but they were also the janitors, (that's how they kept busy at lunch hour) fire-builders and public health inspectors. Every morning we had to line up, show clean fingernails and a clean handkerchief. (It was easy to keep a clean handkerchief in our pockets – we wiped our noses on shirt-sleeves.)

They had a long, slow, sometimes wet, sometimes cold jaunt to get forth and back, and they had to actually teach us hicks a thing or two. Even calling us hicks does a disservice to real hicks – we lived way back behind Hicksville. Most of us hadn't even fallen off the turnip truck yet – we did not act – we reacted.

Take planes, for instance. Airplanes were seldom seen in our skies but when one flew over and if we heard it coming, which we always did, the whole doggone class bolted outside. Summer or winter we bolted and the teacher, caught off-guard, reacted much slower than us. The last little kid would be looking skyward before Teach hollered, and we stayed until the plane was a far-off speck. By this time, she was on the steps, out of breath, out of holler and just stood there as we straggled back to class. They never chewed us out after, but of you peeked under your eyebrows you could see her studying on us, wondering which planet had spawned us.

They never learned, either. If another plane came by (happened seldom) the swift escape-slow recovery was repeated. It's a good thing they didn't try to head us off – they would have been trampled in the stampede.

That first winter of '46 – '47 I learned why the desks were free range. When it turned really cold, the school, uninsulated and with single pane windows, became an icebox. Teacher and Robert lit the barrel stove when they arrived and the desks were pulled into a circle around the stove – littlest kids on the inside, and the older ones in an outside concentric ring. In December, January and February they <u>might</u> sometimes be moved back at noon. We never needed a weather report. When the desks stayed in the line-up for a full day, we knew spring had arrived.

Each double desk had two bottles of Scripto ink that fit into little holes in front of you and your partner. One round only been purchased by the school board, no doubt a few years before, and since then whenever the bottle was drawn down to half it was diluted with water. By the second dilution it was a pale blue. Winter-time was an ink saver – the bottles froze in November and thawed in April.

We practised penmanship with old straight-nibbed pens supplied by the school board. We dipped them into the Scripto bottles and wrote letters of the alphabet from one side of our scribblers to the other – one page for each capital letter from A to Z followed by another 26 pages of lower case letters.

"Calligraphy" time was quiet work – you had to keep the nib from hanging up on the paper, had to keep dipping the pen into the ink, and you had to hold your enthusiasm (what enthusiasm?) in check. Too much ink left a blot, and although each handwriting scribbler came with a sheet of blotting paper, you didn't want to waste it on ink blots – it was much more valuable as spitball material.

If your family had an extra buck or two you might have a fountain pen – usually a Waterman. They had a little rubber bladder inside and a little lever which when operated properly, filled the bladder with ink. Any boy with a Waterman pen soon learned that they had a dual purpose – point it at a girl, pull the lever and ruin a white blouse. It was a good thing those pens didn't last long before the bladder broke and then they were only good for dipping.

By the third dilution of the Scripto bottles the ink was too pale to use, so we practised handwriting in pencil.

One final note re handwriting exercises: it was boring and us boys hated it. The teacher knew this and thus it became punishment if we acted up.

The school curriculum followed the National standard – the three Rs with some geography tossed into the mix. Everyone of our age group will remember the "Hilroy" scribblers with their full page of useful info on the back cover. This was mostly important stuff – inches to feet to yards, ounces to pints to quarts to gallons, and ounces to pounds, for instance. Some info such as chains and furlongs, meant nothing to us. One thing for sure – it was all in Imperial measurements with no liters, kilometers or kilograms to confuse us. We were also taught how to calculate property tax using the "mill rate" concept. I didn't understand it then and I don't understand it now. (Also, for some reason, we were taught to make change using hypothetical dollars and cents. Very hypothetical – to us a quarter was Big Money.)

The real important stuff we learned on our own, on the farm and around the supper table. We learned about animal husbandry, gestation periods, and how every herd of cows had at least one escape artist willing to put up with a barb-wire scratch to steal some oats or barley. We learned that cute little pigs would grow up to be slaughtered – a fact of farm life. The toughest lesson for us youngsters was the annual orphaned lamb. It was brought to the house, wrapped in an old towel and put in a little box beside the kitchen stove. We helped bottle feed it and it became a loved and loving pet. When it rejoined the flock, however, we knew of its eventual fate and it was a tear-worthy goodbye.

Botany classes ran from early spring to late fall. Crocuses were followed by delicate violets, dandelions and bluebells, with later summer daisies, brown-eyed susans and indian paint-brushes. A ladyslipper was a rare find and had to be picked to take home to Mom. Wild roses grew along the fence line on our way to school but were not picked – too prickly.

Biology – another pre- and post-school day lesson. The melting snow filled the ditches along our short walk to school and by early May frogs serenaded us along the way. They interrupted their songs 'til we passed by, and picked up the tune behind us. Soon there would be tadpoles to study on and we watched them grow little legs and lose their tails until they finally made the transition from water-breathers to dry-land hoppers. We would often gently pick up a green frog for an eye-to-eye closer look, but toads were left alone. We knew for a fact that toads would leave warts behind.

It was Christmas Concert time – how exciting! We had been prepping for a couple of weeks practising off-key carols, Christmas poems and a Christmas play. We cut letters from a donated roll of heavy paper for our Grande Finale – all of us on stage in a row, smallest angling up to tallest, holding a message – Merry Xmas. It had to be Xmas – there weren't enough students to spell Christmas.

The day of concert evening older folks built our stage. The teacher's desk was moved to the back and 2x6 planks were spiked onto poplar logs. We even had a curtain! A length of binder twine held donated bed sheets clothes-pinned to the twine. It worked pretty darn good, too.

An evergreen came from somewhere, and after the stage hands left we decorated it with stuff we had made. The school board had found a little extra cash to buy a roll or two of green and red crepe paper and we twisted streamers to string over the stage. Before we left we gave the barrel stove a hefty poplar transfusion – we all would soon return carrying kerosene lamps and lanterns.

Supper didn't last long. I don't know if other kids ate, but I sure didn't – I was pumped!

Finally, all spiffed up with clean fingernails and with spit on a cowlick we headed back to the school. People were coming from miles around - some by horse and sleigh or a single horse pulling a long toboggan, some in vehicles with good mud/snow tires (if the snow wasn't too deep) and some, like us, walked.

There must have been a hundred people there. All the desks were full, chairs (from the Legion Hall in Eriksdale?) lined the outside wall and at the back it was standing room only. The stove needed no extra wood now – body heat and beaming faces warmed us kids to the bottom of our little hearts.

The concert was a resounding success. When the applause died down Santa Claus came jingling in with ho-ho-hos and poor little Richard Monkman screamed and ducked under his mother's dress, staying there until Santa left the building.

Each kid got one present, even the poorest. I know the parents bought something for their own kids, but before the concert started I had noticed someone counting presents. If a kid was missing a gift that someone went outside and returned to quietly add one to the pile. The presents were not ostentatious, either. Nobody wanted to show off to the poorer families – these were kind folks.

We all got a little brown bag with one Christmas orange and five or six curly candies. (school board supplied) and then it was home to bed. As I drifted happily off to sleep I wished Christmas concert time could be a monthly ritual.

Table Scrap

In 1980 I was a single parent living in Winnipeg and one day I took my two daughters (12 and 14) up #6 to show them where their dad had lived and gone to school.

The sideroad was straight now with proper ditches and gravel. The school was gone but the yard was still there with only the odd clump of little willows and the grass was short – it was a pasture these days. The building foundation outline was still visible and I was shocked! In my memory that school was much larger, but if a garage of the same dimensions was erected there it would barely hold my Chev station wagon – my girls were unimpressed.

I walked north to see if our log cabin class project was still standing but I could find no evidence that it had ever existed. The walls had probably supplemented the woodpile before the school closed forever. Even the stumps were unfamiliar – the mature trees had been cut long ago.

I went back to stand in the middle of the school and slowly turned 360 degrees. I could see it all – the stove, the desks, the teachers, and my school-mates. I could hear straight-nibbed pens being dipped in watered-down Scripto ink and scratching across scribblers as we practised our penmanship. I could hear Richard Monkman trying, unsuccessfully, to say "Pacific." (He insisted it was "specific.") Outside I could hear ground squirrels planning escape strategies and was that a plane coming closer?

I wondered – did we have dreams of a better life? I remembered my own dream – it was simple, yet unattainable. I just wanted Mom and Dad to quit fighting all the time.

The other kids were long gone now. Robert was an Air Force guy now living in Nova Scotia. Reg had a career as a customs officer and was in BC. Lorraine Richaud had married a successful farmer near Selkirk and Bert Lowrey was with Manitoba Hydro. Even "Specific" Richard Monkman had done well as an up-and-coming pro boxer until he lost all the fingers of his right hand in an industrial accident.

I don't think we dreamed at all in that school. It was a happy place – a place among friends – a place where our imaginations and plans flourished and ran free, far from the trials and tribulations of home life and unfettered by micro-managed class schedules.

It was a bitter-sweet day for me and I left feeling a bit sad. Perhaps I should not have made that visit, but over the years I've decided that I am glad I did. That little Deerhorn school still has a special memory nook in my mind.

But before we leave the school I've got to tell you about the Lowerys, because although they were one of the poorest families around, I'm sure they were the happiest.

Bill Lowery was a WWI vet on a disability pension. He had left an eye and part of one leg over there and, as nowdays, vets were not treated well by our fat cats in Ottawa. Mothers got a three dollars per month allowance for each kid under 16 and I'm sure Mrs. Lowery's family allowance check equalled Bill's pension. There were at least six kids in the clan and if one Lowery graduated there was always a mini-Lowery to fill the gap.

Mr. Lowery was all rawhide and gristle and I was a little afraid of him. He was never mean, but he never smiled. The thing was - I could seldom look at him face-to-face – that glass eye never moved and I was sure that it was the good one – boring past my "good kid" front to search out my inner bad kid.

Bill had a Model T four-door touring. It was well past its best-before date and the top was long gone, but it ran and how it ran amazed me. Occasionally he would drive the clan to school, picking up Richauds or whoever along the way. Even the non-school age Lowerys came along and everyone would pile out. The puppies ran around laughing until Bill herded them back into the T, which had stalled – and I learned something new.

The Ts had Henry Ford's version of an automatic transmission and when they got old they lost neutral gear, thus when you started it with the crank you stood a chance of being run over.

Bill carried a six-foot pole and two chunks of wood in the car. One chunk was the fulcrum, and the pole lifted one axle. Bert then put the other chunk under one back axle.

Now Bill set the spark and throttle, spun the crank (once,) the motor started and one rear wheel spun in air. Then Bert pushed the car off its stand and with a little spit of gravel the T was off. Bert grabbed the chunk of wood, caught up to the car and tossed the chunk in the back. If Bert was going along, he had to jump in also.

The Lowerys lived off the land. Their shack was on a piece of CNR property not far south of the Deerhorn non-station. They were squatters but they didn't bother CN and CN didn't bother them.

Every summer all the kids picked wild strawberries, raspberries and saskatoons. In the fall, like us, they harvested hazelnuts. On school weekends they helped Bill cut firewood and dig snake-root (Seneca.) Snake-root, when cleaned and air dried, fetched three or four or more dollars per pound.

One Saturday in spring or early summer I got to go to the Red-And-White store with Dad or Walter. While store and post office stuff was being done I chugged on down to visit the Lowerys. Mrs. Lowery answered the door and she had no angles whatsoever – she was round. Her body was round, her arms were round, her face was round and she smiled and laughed as she walked around. I got a round hug, a round laugh and was invited to share a round of jam sandwiches. If a little Lowery came in crying from a stubbed toe it got a round hug and laugh and soon carried both back to the yard.

The Mrs. Lowery disease was infectious – how could anyone ever be unhappy in that house? And believe me – I never called that house a shack again.

(Meanwhile back at the farm, my childhood existence was a mixed bag of good and bad times, which I thought was uniquely my cross to bear, never thinking that other kids had their own ups and downs.)

That first fall of '46 we had no garden to harvest but we picked hazelnuts – a gunny sack full. Then we husked them and ended up with two gallons of nuts which, when combined with a couple of small bags of walnuts and such at Christmas lasted well into spring. We had seldom picked hazelnuts in New Brunswick because their hulls were covered with tiny spears, but for some strange reason the spines disappeared at the Man/Ont boundary and the nuts were husked painlessly.

I was seven going on eight now, and my chore list was growing faster than me. Once Dad had some confidence that I wouldn't get lost myself I replaced Rod on cow-fetch duty. In the early spring up to about mid-June they wouldn't go far, but as the summer heat dried the open pastures, they strayed into crown bush pasture. By fall the buggers might be three miles away and me and Buddy sometimes had a tough search. Two cows carried bells, but what really irked me was that when they were finally rousted out it was past regular milking time and their bags were uncomfortably full. Now they led the herd home lickety-split leaving me far behind. Why didn't those dummies figure that out two hours ago?

We had a new Buddy now. Buddy One had failed to stop, look and listen and Rod had buried him back in New Brunswick. Buddy Two was a female, a Shepherd/Collie cross and she came with the farm. She was a wonderful dog and bonded immediately with her new family.

And could she fetch cattle! In early summer, if the herd was in sight, all you had to do was say, "Fetch!" and fetch she did – all by herself. She never ran the cows but when a nip was deserved, they got one. They had a mutual respect for one another.

Sometimes the cattle were too far away to the northwest for Buddy to see. If I thought they might be in that direction, I would go to the back of the house where a ladder led to the roof of the kitchen addition. Another led me to the house peak and since the square house had a four-slope roof there was a two-foot by three-foot flat top by the chimney. Aha! There were the cows, almost a mile away.

I had a brain wave – why not teach Buddy to climb ladders? I took her to the ladder but that old dog was not about to learn that new trick, so off we went to bring the cows in, Buddy as usual trotting beside me proudly waving her tail.

It may have been 1948: One beautiful spring morning Buddy laid down in the corner of the front yard where our driveway curved toward the house. She watched the sunrise, thinking of her puppy-hood days and drifted off to a permanent sleep. Mom found her there and they buried Buddy on that spot. Judy and I were told when we came home for lunch – what a sad day. Buddy had no headstone but perennial flowers were planted on her grave. Good dog, girl – you'll never be forgotten.

Mickey, a Border Collie, replaced Buddy, but he had to be trained, and Dad would do that himself. If I trained him he might pick up some bad habits. Well, gee whiz – were my habits so bad you didn't want a dog to catch them? So I went out dogless and it was even more frustrating. Without a dog's superior hearing and nose the cows virtually disappeared.

We had a riding horse, Babe, who had also come with the farm, and it was suggested that I ride her to fetch cows, as Rod had done. We had a saddle, but the stirrups could not be shortened to my leg-length so it was bareback for me. I wasn't fussy about it and neither was Babe. We made it less than a ¼ mile down a bush trail until Babe found what she was looking for – a low hanging tree branch. She ducked her head and I bailed out, smart enough to hang onto the reins. Now I had no one to boost me up to remount, so I spied a large rock. I stood on the rock, and Babe and I faced off. I could stand on the rock forever if I felt like it but she would not stand near enough for me to jump on. I eventually found the cows, leading Babe all the way. It was a bummer for me but no big deal for her as long as she got to grab the odd chomp of nice-looking grass. After that day I walked alone.

Babe was a crafty old girl – one of our neighbours said she was at least 22 and, like the evil axis interrogator, she had "Vays to make make you valk." I can remember well the first time Walter saddled her up.

We had a horse barn with stalls for four horses and it had a single door. Walter brought Babe out, cinched up the saddle and climbed aboard. Babe bucked a little – just a couple of crow-hops before deciding that they might as well head out. Down the road they went.

It was maybe fifteen minutes later that they came back with Babe galloping flat out and Walter holding one rein! The barn door was open and Babe never let up on the gas pedal. Walter bailed out at the last second and did a face plant on the outside barn wall. It was either that or lose his noggin on the door frame.

Walter told us that s soon as they were out of sight Babe gave her head a shake and snatched one rein out of his hand. He tried to snub her up with the other rein, but she kept circling until they both got dizzy. He had only one recourse – he gave her back her head.

Table Scrap

Our dad bought an Angus cow and Walter was sent to fetch her with the two-ton. Dad knew all about the Angus temperament – Walter not so much.

It was wet and muddy in the farmer's yard and they could not get the truck to the loading ramp. A make-shift ramp was set up, and with much difficulty they got a rope around her neck. With two men on the rope they brought her to the ramp. She didn't like the looks of this deal and chose freedom. The other guy bailed out but Walter was as stubborn as the cow. He got a wrap on a tree and snubbed her up.

She studied the situation for a bit, figured out the difference between clockwise and counter-clockwise and started to chase him around the tree. With every revolution her end of the rope got longer and Walter's got shorter until, to save his fingers, he had to let go. She headed for the pasture, and I don't know if they ever got the rope back.

When the farmyards dried up they did manage to load her and bring her home. It's a good thing our stock rack was strong and had six-foot walls – she kicked the dickens out of it. At home no ramp was needed – they opened the tailgate and let 'er rip.

So she joined the herd – sort of – she wasn't much of a team player. (Our herd was dual purpose – we shipped cream and beef.) She dropped a couple of good calves when and where she felt like it, sharing her milk only with her own offspring. She never hit the Friday night after-milking dairy bar but often joined in the Saturday stampedes – usually as leader. When we moved to the dairy farm in the Rainy River Valley she was dropped off at the stockyards in Winnipeg.

The poem that follows was one of Mom's prize entries in her scrap book:

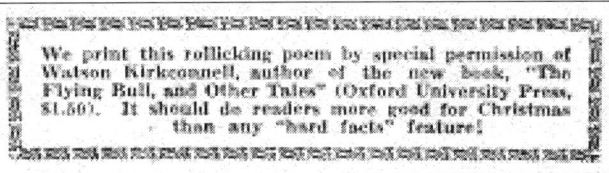

I

The toughest bull I ever saw
    Was on a farm near Neepawa,
    Where an old cowhand, Dave MacMeans,
Kept a big herd of Aberdeens.
Dave was a rough and wrinkled Scot,
With bulbous nose and sunset whiskers;
He liked his whiskey neat and hot –
When young he'd frisked among the friskers.
And sixty years had left him cursed
With stomach ulcers and a thirst.

Two things he'd loved in life's long battle:
    Theology and Angus cattle.

And while the Calvinist in him
Waxed fervent on predestination,
He'd argue long with equal vim
That black bulls were the farm's salvation.

The pride of his own dusky herd
    Was Mumbo-Jumbo, bull supreme –
    The biggest, grimmest, blackest-furred
Of all the brutes of Pharoah's dream.
He was as black as Satan's dam,
And nigh as tall as Pilot Mound;
The rumblings of his diaphragm
Made thunder thirty miles around.

His ribs were like a Roman Arch,
    His back was level as the prairie.

His massive legs in stately march
Made a small earthquake through the dairy.
But of his eyes no words can tell –
Within them burned the fires of hell,
Two lamps of livid yellow, lit
With Anger from the nether pit.

His cows had always found him kind,
But hatred smouldered in his mind
For all our human Jacks and Queens
Except his Master, Dave MacMeans.
Yes, Dave he loved beyond a doubt,
Because the old sanctimonious sinner
Would often give his mammoth snout
A snort of whiskey after dinner:

And so a spirituous bond
Kept beast and man uncommon fond.

II

Now, in the spring of '35
David was gathered to his fathers.
One Sunday he had been alive;
On Monday, fierce internal pothers
Brought on that final, fatal quiver
That end cirrhosis of the liver;
And so by Wednesday night he lay
Dead sober in the graveyard clay.
Then distant heirs and lawyers, dark as
The vultures at Gehenna's gate
Came flocking in to share the carcass
Of Dave's unfortunate estate.

Farming they held in cold derision;
So, to facilitate division
And settle up with one clean slash,
They auctioned everything for cash.
One August morning, hot and clear,
A leather-larynx'd auctioneer
Stood in Dave's farm-yard on a table,
Half-way between the house and stable;
And there, amid a throng of buyers,
He bawled the merits of each chattel,
From combines down to common pliers,
And last, not least, Dave's Angus cattle.

But offers were not plentiful
For Dave's notorious black bull,
Whose most unmitigated choler
Was reckoned dear at half a dollar.
In vain the auctioneer avowed
That any farmer might be proud
To own so vast a thoroughbred
Most famous in his progeny –
For, looking in his eye with dread,
Each thought he'd let the Monster be.

Just when it seemed the day would end
Without a bid for the old devil,
One quiet voice agreed to spend
A hundred dollars, on the level.
And thus was sold the brute unruly
To Deacon Williams of Plum Coulee.
A man as good as he was strong
And pious as the day is long.

But when at last each buyer sought
To drive away what he had bought,
A streak of dark Satanic strife

In Mumbo-Jumbo came to life.
He pawed the ground, he snorted fire,
And one could see him in his ire
Fiercely and visibly determine
To rid the farm of human vermin.

Straight at the human throng he charged,
Straight at these puny things of shame:
A frantic lane of fear enlarged
To leave him passage as he came.
Over the fences did they leap,
Grasshopper-like by tens and scores;
Shunning the bull's destructive sweep,
Awed by the bull's appalling roars.

Against such cars as had been left
Venting on them his two-ton heft,
In efforts to set the world aright –
A thunderbolt of Hate and Night.
A dozen Fords were overturned;
He tore the fenders off ten Nashes;
Sparks from his onset lit and burned
A score of Pontiacs to ashes.
When nightfall closed the day's wild session
It found the bull in full possession.

But hours later, far from thence
Good Deacon Williams, sad but trusting,
Invoked the aid of Providence
To give the Devil's bull a dusting
"Humble his spirit to the earth!
Give me my hundred dollars worth!"
Then with an "Amen" loud and deep
In simple faith he turned to sleep.

III

After a night of breathless heat
There dawned the hottest day as any
Had ever known. Rays seemed to beat
As from a vaster sun; and many
Thought of that final day of ire
When all should be destroyed by fire.

Small wonder was it no one went
   On such a day to Dave's old farm
   To see the bull's dark discontent
Or seek his frenzy to disarm.
Parched but triumphant, hour by hour,
He stood there in insensate power.
Alone, unchallenged, black as ink,
He scorned to bellow for a drink.

Two hours past the gaping noon,
   A dark cloud rose to west-northwest –
   Slowly above a world a-swoon
It reared with thunder in its breast,
A roaring, swirling, cloudy funnel,
Black as the entrails of a tunnel.
But though the twister nearer swept,
The Angus bull remained defiant;
Dauntless he stood to intercept
The black, intruding, cloudy giant.
But all in vain: its mighty force,
Seizing him swiftly from the ground.
Propelled him on a skyward course –
A heavenly bull, and southward bound.

And with him went the shattered hulk
Of Dave's best barn; and sped from sight
Revolving 'round his darker bulk
Like some infernal satellite.

IV

An hour later by the clock
The storm near Deacon Williams passed,
So close to all his barns and stock
That the good brother stood aghast.
Then, mid the tumult of the storm
He saw a black Satanic form
Swoop down, as though on hidden wings
And light upon the prairie clay,
While with diminished thunderings,
The great tornado ebbed away.

Not fifty yards from Williams' door
There lay a miry open slough.
From it now came a mighty roar!
Out rushed the Deacon, swift to view;
And there, by heck, Stood Mumbo-Jumbo,
Up to his belly in the gumbo.
By "Act of God" delivered duly
To his new owner in Plum Coulee.

What thoughts had thronged his heavy mind
During that epoch-making flight
No one can tell: but I'm inclined
To think he got a thorough fright.
For, all the rest of his black life
He was most mild – most timid, maybe –
And often Deacon William' wife
Would leave that bull to mind the baby.

In the summer of 1947 we were a four-horse, tractor-less farm. Dick, a medium-sized black came from I know-not-where. Bob and Queen had been bought when we moved in last fall and they were immediately re-christened Dan and Bess. Dad's first name was Robert although he had always gone by Wilbur, his middle name, and Mom's name was Queenie - so you can understand why the horses' names were changed.

Dad tried to tractor up. He bought two old Rumley Oil-Pulls, one for spare parts. The Oil-Pulls were famous for their reliability in his younger days but now they were old and cranky. They never did do a lick of work and soon went to the scrap yard at three cents/lb. They each weighed almost two tons, so there must have been some profit.

Dad, Walter and Rod put in 60 acres of oats with the horses and because the land had lain fallow for a couple of years they took off a good crop.

Now Dad borrowed a horse binder to cut the oats. They were stooked to wait for the threshing machine which came up the road one morning in September as I was walking to school.

I had seen threshing machines in farmyards and farm papers, but the tractor pulling it was something else. I found out that evening that it was a Field Marshall, made in England – and what a weird duck to cross the pond to Canada.! It was close-coupled, green with red wheels and had a high exhaust stack. It chuffed by me, jiggling and shaking, and continued on farmwards. I could hardly wait for four o'clock.

They were still threshing when I got home and I enjoyed the new experience. Neighbours' wagons brought in loads of sheaves, the separator swallowed them, straw blew onto a continually growing stack, and oats flowed into our two-ton. Throughout all this activity, the Field Marshall, now on drive belt duty, chuffed calmly, still jiggling and shaking.

True to English form, the Field Marshall had the belt pulley on the left.

The threshing was finished before we returned from school the next day. We had heard the tractor and threshing machine passing our school, but I was eager to get home anyway.

First I checked out the straw-stack and it looked like a mountain. It was the only hill I'd seen since we hit the prairies and I filed it under "Flexible Flyer Opportunity." I went to the granary, now full of pale yellow oats. They smelled good and felt good when I picked up handfuls and let them trickle through my fingers. I decided that harvest-time was next to Christmas on my "like" list.

(Our last threshing bee would be at Crozier in 1953 with John DeGagne's threshing machine and John Deere tractor power. I was allowed to join the spike pitchers loading stooks of sheaves onto wagons. I was fourteen, but now I ate with the adults.)

Supper that night, as I knew it would be, was an informative family conversation. Among other things I learned about Joe Richaud and the Field Marshall.

Mr. and Mrs. Richaud had a good farm just this side of Deerhorn and their two kids Guy and Lorraine went to school with us. Joe had fought in WWII and had become infatuated with English mechanical stuff. (Dare I say – junk?) When he mustered out he bought a Thames three-ton grain truck and the Field Marshall tractor. The Thames was hopeless. It hauled less grain than our Ford two-ton, would never start in cold weather, and if a part was needed it came on a slow-boat from England.

The tractor was slightly better than the truck. It had infinite drawbar and belt pulley power and was easy on fuel. It wouldn't start in cold weather either, but Joe didn't need it in winter. I was told it had one cylinder – a big one – and that was why it jiggled and danced. The transmission was bidirectional with three speeds forward and three reverse. This was because you never knew which way the piston would head for when you hit the shotgun shell. No kidding!

You could crank the huge flywheel to roll her over but that was hard, often futile work. If you rolled the piston up to top dead centre, put a thing shaped almost exactly like a twelve-gauge shell into an aperture on the cylinder head, stood back, shielded your eyes and hit the firing pin with a hammer, the F-M would come to life. This was hard for me to believe, but was actually fairly accurate.

One wild story, which I heard later, went this way.

Joe had been cranking the F-M, got tired and went the shotgun route forgetting the crank rod was still in the flywheel. When he fired the shot the flywheel spit out the rod which impaled the barn wall. For years I thought this was a true story until I found out that a cranking rod did not exist.

In 1948 Joe joined the John Deere club, the F-M was used only for threshing and the Thames slowly sunk in the mud waiting for a part that never came – Joe was de-anglicized.

Manitoba winters, though cold, had their good points. Winds made snowdrifts along the fence lines and then sculpted them into beautiful, flowing works of art. When a blizzard created a huge drift by our little shop, Judy, Frank and I made a snowbank hotel with inter-connecting rooms. We sat, pretending to be Eskimos, but soon went back to civilization by the living room stove.

Me 'n Judy and the Ford, all three of us dressed for winter.

Evenings were the best. We sat at the living/dining room table and read or played board games by the light of a softly-hissing double-mantled gas lamp. Mom and Dad seldom argued in the evenings so it was extra nice. The radio was turned on (always tuned to CBC) and after Matthew Halton read the news and after the farm report was over we actually got to listen to half-decent stuff. It might be Amos and Andy, Our Miss Brooks, or Fibber McGee and Molly. Then it would be time for music, we would lose interest and I would go back to my book or the Saturday Evening Post. It wasn't that I didn't like music, but Dad, always the Captain, controlled the knobs and thus the airwaves. No country/western for him – CBC played classical stuff only. Sometimes the fat lady screamed and I perked up. Maybe she'd been goosed by a fiddle bow, but she soon settled down and I went back to my reading.

Saturday night was Hockey Night in Canada and was never missed. "From the Gondola high above Maple Leaf Gardens, this is Foster Hewitt and this is Hockey Night in Canada." I had a picture of that gondola in my mind, and many years later a TV documentary about the Gardens (before it was torn down) surprised the hell out of me. The Gondola was exactly as I saw it in the '40s and I do believe that Foster had to crawl along a rafter to get to it.

What heady hockey memories – Syl Apps, Turk Broda, Rocket Richard, Elmer Lach, and the three Bentleys from Saskatchewan – Max being the best. I loved them all and although Toronto was my team, I cut them all from the paper and put them in my scrap-book. Does anyone remember Sweeny Schreiner? Well I do – in 1943 he scored the hat trick when Toronto beat Detroit 3 – 0 to take the Stanley Cup.

I had Metro Prystai (Chicago) Gus Mortson, Edgar LaPrade (Montreal) and I had the complete write-up of Billy Mosienko's 21-second hat trick – but sad to say, I no longer have that scrap-book.

Our radio was a cathedral-style Deforest-Crossley or Marconi. It was "no touch" for us kids. It ran on three different sizes of dry cell battery packs and they were expensive (too expensive – often repeated in dealing with everything from radio batteries to Daisy Genuine Red Ryder Lever-Action BB Guns.) Sometimes, right after school, if Mom was in a good mood, and if Dad was not around we could turn it on, but even then there were restrictions. We could listen to the Lone Ranger, Yukon King (he was a great dog) or Flying Doctor. This was an Australian show about a doctor who flew his own plane around the Aussie Outback. (Come in, Flying Dowctah, come in, Flying Dowctah, where are you, Flying Dowctah?) After our radio time was over we had to put the dial back to CBC so Dad wouldn't catch on.

Jealousy and envy were always part of my existence. Reg and Robert had battery radios at home, but theirs were powered by six-volt car batteries, and they had wind chargers to keep the batteries up to snuff. As long as the winds blew they could listen to the radio whenever they wanted. Every day at school I had to listen to them rehash last night's Charlie Chan, Boston Blackie, or the Great Guildersleeve, and these were programs that for some unknown reason were unfit for our ears. I'd counter with the Lone Ranger with some success, but our other listening pleasure was "kid stuff" so I eventually hit my own mute button.

In those days, as some of you will remember, the NHL Playoffs actually ended before summertime. We didn't need hockey now, anyway. There were new calves in the calf pen, the first spring lambs were arriving and the two sows, who had lived very comfortably all winter in the straw stack had obviously farrowed. We hadn't seen the piglets yet, but soon they would be rousted out to be moved with their mothers to the pigpens. There was plenty for us little kids to do and thanks to ever-lengthening days plenty of time to do it.

Spades, garden hoes and rakes came out of the shop. It was our job to keep little rivers open to drain snow-melt water. Special attention was paid to the house well. Cow poop had built up last winter and as ice and poop melted it had to be raked away from the pump. Mom's water must be kept clean – the barn well did not need protection.

When we knew there was little chance of a cold snap the rain barrel was put under the downspout. The first good spring rain would fill the barrel and our hearts – winter was over!

Sidebar: There was a barn well and a house well. Both were hand dug shallow wells, perhaps twenty-five to thirty feet deep. If it had rained a lot before freeze-up the barn well might last the winter, but last fall had been a dry one. By mid-January a second stock trough was set up beside the house well and by mid-February both were running short of water. Now the cattle were watered in shifts – four at a time at each well with breaks to allow the well-water to recover somewhat. The process started in the morning and would go on all day long. The last horse got his drink just before we had supper.

In early May Dad and Mom went to Lundar and came back with little boxes of garden seeds and a big bag of seed potatoes. Judy and I helped Mom plant the radishes, corn, peas and such, and as a reward we got to put in a row of peas for us kids. I've always loved peas straight from the garden. By late May we had fresh radishes and lettuce. Peas followed later, and by early August the potatoes were in full bloom, the corn was tassling and the cabbages, which had been dusted weekly with "paris green," were worm-free and getting bigger.

And then – disaster! We got up one mid-August morning to find the garden trashed. The corn, the cabbages and the whole deal was a trampled mess. The cows had raided the joint, leaving cow-pie calling cards in their wake.

Mom, who had been pretty calm all summer, went ballistic and we little ones headed for our hidey-holes. Dad fixed the fence but it was too little, too late. In September we were able to dig some carrots and potatoes. The cows wouldn't eat the tops and quite a few of those plants survived the walk-about, but the corn and the cabbage had been eaten.

The following year was a repeat performance. The thing was – the garden fence was not permanent. It had to be taken out every fall so the garden could be tilled because in those days the tiller was a team pulling a sulkey plow. The same team pulled the double disk over the garden in the spring and the fence was re-erected, thus it was never that well built. The garden was on the south side of our driveway in the corner of the night pasture and by the middle of every August, when pasture grass started to get scarce, the garden was raided again. It was a family tradition.

School was out, haying was not yet underway and Rod built me a ship! We had a decent dug-out, at least 50' x 120' just to the south of the shop – this would be the "Broad Pacific". He found a 20-inch piece of 2x4 and shaped one end for the prow. Then he hand-bored two shallow 5/8" holes on the centreline – this would be a two-master brigantine. He now drilled a 3/8" hole through the 2x4 near the stern to hold the rudder shaft. The he turned the hull upside-down and with a hand saw cut a length-ways slot dead centre in the bottom. He found a piece of flat steel about 3"x4" and tapped it into the slot. This would be a double-duty deal, acting as a keel and also as ballast to hold the ship upright.

With the basic hull in the shop-vise drydock, he got to the important stuff – masts, sails and rudder. I was helping by holding things when necessary, but mainly by keeping my mouth shut. I didn't have to ask many questions because Rod explained everything as the ship-building progressed.

The masts went in, and spars and cross-arms were carefully rabbit-wired to the masts. All were straight pieces of willow – we had no wood dowelling. Little nails (Rod called them "brads") were tapped into the side of the hull. Strings led from these to the ends of the spars/crossarms to hold them in place. Writing-paper sails, three on each mast, were attached, I no longer remember how. A triangular paper jib was now held by another string which ran from the top of the foremast to the bowsprit (another little willow stick nailed to the bow.)

Last but not least the rudder was installed which was another willow stick fitted through the hole in the stern. The bottom end was split, a square piece of tin inserted and the split end was tightly rabbit-wired so the tin rudder would stay in place. At the top a tiller was attached using willow and rabbit wire again. Now Rod tapped in brads in a semi-circle near the rudder shaft, telling me this was to set the tiller amidships, port or starboard, to allow the ship to tack.

What was he talking about? He'd been tacking stuff on for days!

LAUNCH TIME! The build had taken three days (Rod had his own farm chores to do) but one nice day with a proper light breeze from the north, the champagne bottle was broken across her bow. Off she went, perfectly upright, perfectly balanced aft and forward, and she even kicked up a little bow wave. Marvellous!!

We went to the south end of the dugout, picked her up and put her back into the water just up the west shoreline. Rod set her rudder, gave her a little push and the wind heeled her over until the bottom of the starboard sails were just above the water. The rudder brought her into the wind and she paused momentarily until the wind caught the jib and turned her bow to the right. Now the sails filled and she did another tack, and with one of us on each side of the Pacific she tacked all the way home.

With Judy on one side of the pond and me on the other we sailed her every chance we had (Frank could watch, but not touch) until inevitably we got careless and left her on shore, expecting to return after supper. Of course, the cows came for water and the HMS Marvellous was history.

Rod had engineered and built a working work of art and ten years later he would graduate from Western U. in London, Ontario with a civil engineer's degree.

That ship-building experience played a part in my future as well as Rod's. As for me – I simply became a sea-going buff. We had editions of C.S. Forester's Horatio Hornblower in our bookcase and during that fall and winter I read every one. In my evening dreams our little brigantine became a man-o-war and the scuppers ran red with blood as we kicked the poop out of the Spanish Armada. Sometimes she was still a 24-gun brigantine, but we could get so close to the big galleons that their cannonballs flew harmlessly overhead as we pounded them at the waterline until they sank.

After Horatio had been read and re-read I found a paperback – The Golden Hawk. It was about a swashbuckling semi-pirate. That book had a relatively harmless section wherein the Hawk and his beautiful lady hostage did some canoodling and her bosom heaved. I guess I marked that page too well and the book disappeared.

As for Rod – he claims to have few memories of our ship, but I have a theory: That attention to detail he showed at such a young age was an indication of how he would handle his future career which, after graduation, was with the RCAF. It was an exemplary, varied, sometimes exciting and dangerous career path.

Among other things too numerous to mention or recall, he took a test pilot course at Edwards AFB in California, and after completing the course stayed on for two more years to instruct fellow pilots. Among his students was Michael Collins, pilot of the lunar command module that circled patiently while Neil Armstrong took that famous step.

One day, Rod walked into the ready room and Chuck Yeager asked him to go for a spin. It was in a two-seater and Rod flew the front seat while Chuck did some USAF stuff in the rear. Rod said that Yeager was a down-to-earth nice guy – unimpressed with his own fame as a test pilot. It was the start of a lengthy relationship – Rod often flew with Col. Yeager after that.

Later on, Rod flew interceptors – CF101 Voodoos out of Val D'Or, P.Q. Two Voodoos and two pilots lived and slept in close contact for four days at a time until the swing shift relieved them.

And when Dad died in 1974, Rod flew a T33 up from Ottawa, landing in International Falls, Minnesota. (No graft here – all RCAF active duty pilots had to maintain X number of flying hours.)

After the funeral we watched Rod leave as did everyone at the airport. He took off southeast into the wind, and disappeared. Some folks standing nearby turned away, but I told them to hang on – he would soon return. (Note: This was set up with the control tower beforehand.)

He suddenly reappeared from the southeast, fifty feet above the runway, waggled his wings, and banked sharply, upward and onward to the east. It was a thrilling and fitting tribute to the Old Captain, and I still get 1974 goose bumps.

I got a real treat in the early spring of '48 – I went to Winnipeg with Dad in the two-ton. It had to be a week-day, which was strange, because we were never allowed to miss a day of school. Perhaps it was a late Easter that year, or maybe it was a birthday present – but for whatever reason, I Was Off to The City!!

Up until the mid '50s, #6 Hwy was all gravel, all washboard, all the time. I'm sure grader operators had to pass a "making washboard" test before being turned loose. New cars and trucks didn't stay new for very long on that road. It loosened every nut and rivet and I was surprised that the ditches weren't full of stray fenders and bumpers. Then we hit pavement at Notre Dame Ave. and like magic, the Ford went silent.

We dropped a couple of cull cows at the Burns meat packing plant on Notre Dame and continued on downtown to J.H. Ashdown's Hardware at Main and Bannatyne. Dad needed something, I went in with him, and that place was huge! Since 1871 Ashdown's had been the main wholesaler to Western Canadian hardware stores. This was their only retail outlet and they had everything a boy could want and never get.

Bicycles! Hundreds – thousands of bicycles were suspended from the ceiling and off into infinity at the rear of the cavern. All brands in all colours of the rainbow were up there with softly gleaming chrome fenders, some protecting "balloon" tires. We had a perfectly good double-barred CCM family bike at home, but I sure wished I could road test those babies.

Then it was off to Hambley's Hatcheries to pick up chicks. (The irony strikes me now – at eight years old I picked up fifty chicks – ten years later I would have trouble picking up one.)

Hambley's was also downtown – everything was in downtown Winnipeg in those days. Dad asked for 50 Barred Rocks, and while the Rocks were picked and boxed we got a hatchery tour.

We walked down an aisle in a hot, humid room with trays of eggs on racks on both sides. This was the incubator room, and as we went along we could see that we were following the incubation process. Soon we passed eggs with little beaks sticking out of little holes in the eggs and the holes got larger and the beaks became heads and before we left that section there were little wet chicks and empty eggshells on the trays. The next room was also warm, but comfortably so. Now the chicks were soft, sweet-smelling fluff balls and little "cheeps" became stronger and louder as we went along. Finally, we reached the section where boy and girl chicks were being placed in separate cardboard tray-like shipping boxes. I watched transfixed as an oriental guy picked up each chick, turned it upside down for a look and placed it in the appropriate box. He handled them gently and so swiftly that it was hard to follow his movements. I was probably told that he was "sexing" the chicks but sex was a word never used at our house – I thought "sex" was a poorly-pronounced number.

We wash-boarded home with the box of chicks between us. They seemed to handle it better than I, with just the occasional "cheep" from the cheap seats. If we had asked them, I'm sure they would have opted for a train ride. Our spring chicks usually arrived in Deerhorn by rail, but I guess today Dad had taken the opportunity to save the two-dollar shipping charge.

Sidebar: The chicks were put into a cozy corner of the chicken house. The next day I went out there and did my own chick-sexing, but as far as I could tell, an upside down chick looked no different than an upside-up one. Many, many years later a TV doc answered the question. Male chicks have a "hook feather" near where their wings attach to their body and this was what the oriental guy was looking for.

Towards the end of May that same year we tractored up! It surely had been under discussion for some time but it came as a complete surprise to me. Dad drove into the yard one day on a brand-spanking new John Deere Model B tractor pulling a new two-bottom plow. What a bright, green-and-yellow red-letter day that was!

She sat in the warm sunlight and cooled down with a quiet "hiss" here and there while we all welcomed her to the farm and family. We stroked her nose, her flanks and gave her as yet un-rock-scarred tires a pat. We all got to sit on the padded (with arms) comfort-controlled adjustable seat (forward and backward) in order of seniority, including Judy and Frank, who had to be lifted way up there. Some tried the lights, or the six-speed gearshift and hand clutch. Walter demonstrated the hydraulics – push the "power-trol" lever forward, and the plow raised. Pull the lever back and the plow dropped. My only thought as I held the steering wheel was that you couldn't lose a rein on this horse.

That night after supper We got to read the owner-operator's manual, again in order of seniority. We youngest had to go to bed before our turn but we slept well – what a day!

I think that tractor and plow set Dad back about $1200, but it was money well spent. The Model B did everything asked of it until it was semi-retired in 1960.

Table Scrap

Dad bought the John Deere from Danielson's Garage in Lundar. In 1999 I dropped in with my brother-in-law Leon. They no longer had the heavy tractor dealership, now handling parts only, and Leon needed some. While he was at the parts counter I mentioned to a nice lady at the office that my Dad had bought a tractor from them 51 years ago and she fetched the current Mr. Danielson, who was now an older gent himself. He went to his office and soon came out carrying a thick binder and in that binder was a record of every John Deere his dad had sold. They were not organized alphabetically, but by year and date – a page of each purchase. I flipped through it until I found: 1947 – Marcel Henrotte – Model A – followed by: 1948 – Robert Wilbur Durnin – Model B, Joe Richaud – Model A, August Richaud – I Model G, and again, in 1949, Marcel Henrotte – Model M.

Some names before and after I recognized, but they had lived beyond my little trapline circle in Deerhorn. What really gave me a chuckle was that every individual page had space provided to record recommended, authorized and guaranteed John Deere periodic service. Not one farmer took the opportunity to haul or drive his tractor back to Lundar to pay for service he could do quite well himself – thank you very much.

I don't know if Danielson's is still in business now. If not, I hope that book has been donated to the local museum.

So now, with a tractor, haying would be easier (?) and faster. First of all the horse sweep was converted. The sweep was sort of a large fork – about twelve feet wide with wooden steel-tipped eight-foot tines that slid on the ground. The back was a four-foot wall of boards that also extended up each side about five feet. This was to cradle the hay as the "bunches" were picked up to be brought to the stacker. A horse was hitched to each side of the sweep to provide two horsepower. The Model B would replace them.

First a frame was made and bolted to the front frame of the B. It had two holes at its front end and these would be the sweep's pivot points. Two angle irons were bolted upright to the back of the sweep, another two were fixed to the "rocker shafts" on the B and cables were run from the top of the sweep irons to the bottom of those on the shafts. Now, with pins holding the sweep to the frame, the hydraulics would pull the sweep teeth off the ground for transport or to clear rocks. It was ready to put up hay.

But first the hay had to be cut. Last year a four-foot horse mower was used. It was as slow as molasses in January so Dad bought a used Massey-Harris trail-type with a six-foot cutter bar – a fifty percent length improvement. It was a pre-war model and there was no steel shortage on this baby's assembly-line. An implement tire at each rear corner held that end up, but it had no integral jack on the drawbar and it was a heavy lift for one man.

It was pto driven – that worked fine and dandy, but it had a manual lift. If a rock was in the cutting bar's future you had to lean way back over the tractor seat, grab a long handle and hope you were in time. If you were too late – which you always were – the bar caught the rock, the spring-loaded hitch released and the B drove on mower-less. It only took a day or two for Dad and Walter to realize this was unacceptable, so the mower was taken to the welding shop in Lundar for a refit.

It was a simple solution: a bracket here, a bracket there, and the hand lift was replaced by a hydraulic cylinder – the same cylinder that operated the plow lift. As an added bonus, the cutting bar lifted much higher now and would clear all but the highest boulders.

There was one fly in the ointment. If you were late on the hydraulic lever the bar caught the rock, the spring release hitch dropped the mower and the hydraulic hoses were ripped off with loss of fluid and an expensive and time-consuming repair was required.

This only happened once and again the problem was easily solved – the spring release became a solid hitch. Now if a boulder was snagged the mower stopped the tractor dead in it's tracks. It was hard on the cutting bar guards, but the strong Massey handled the stress, as did the B.

Sidebar: I should point out that most of our hay was on "parkland." We owned a hay quarter that was "unimproved," meaning it was uncleared, unbroken land. This was wild, native hay, and we cut around bluffs of poplars and willows and up to the slough edges. We'd pick the odd stones that were on the surface, but the larger boulders were difficult to remove. At the home place our cultivated tame hayfield was kept stone and boulder free.

Table Scrap

Everyone took their welding jobs to Balbon's Welding Shop in Lundar. No farmer, nowhere in our #6 neck of the woods had a welder. If you lived more than a half mile off the highway you had no electricity, and even those who had power had no modern compact welding units to buy. Mr. Balbon, of course, had electricity, but he also had a big welder, so he had to power it with a "hit-and-miss" stationary engine.

The hit-and-miss concept is simply a speed control. The engine has no throttle – it runs at one speed. Under load it may fire continuously but when the welder guy moves to another spot the engine goes into "coast" mode and only fires every second or third power stroke. Thus the "putt-putt-putt" becomes "putt-putt-chuff-chuff-putt-putt-chuff."

I went to Lundar with Walter one day and while Mr. Balbon was welding I was chased outside – they didn't trust me to avert my eyes from the super-bright arc. I wandered to the side of the shop and what to my wondering eyes did appear, but a series of smoke rings in the air. The hit-and-miss exhaust pipe protruded through the side of the shop wall and with every putt, a perfectly shaped smoke ring wafted into the air. It was dead calm that day and the rings enlarged to about a six-inch diameter as they rose, before they were dispersed in the leaves of a nearby tree.

Think about it. Nowdays a nine-year-old kid would take a cell phone picture, but I think mind pictures are far superior. I have clear, unfaded recollections of that day – the buzz of the arc, the blue/white flash reflecting off the open shop door, the pleasant aroma of hot steel and those perfect smoke rings still waft through my memories.

A couple of days after a field was mowed, the hay was raked and "bunched" with a two-horse sulkey rake. This was not rocket science – the first pass left the hay in long, windrows, then the rake went down the windrows bunching up small mini-stacks. When enough hay was ready we shifted gears into the stack-building mode.

We used an "overshot" stacker as did all the farmers thereabouts. It was basically a gin pole arrangement with some modifications, all made of lumber and quite heavy. It was moved (skidded) by horses or tractor to each new stack location.

Note the two-horsepower sweep on the left-hand side of this typical forties hay-time scene.

The B did triple-duty now. The sweep was attached to bring the bunches to the stacker and when that stack was finished the sweep was dropped and the mower was reattached to knock down more hay. Then while the horses raked the hay, the B moved the stacker and the sweep went back on.

We were a three-horse farm now. Old Dick had died and now Bess and Dan pulled the rake. Little Babe was on light duty – her only job was on the stacker rope. So by guess and by golly the hay was stacked by late August, with men, tractor and horses resting impatiently when it rained. Dad would fret about things – was there enough hay this year? Was there any quality in the late-August hay? (There was always late-August hay because the wild hay was too short to cut before then.)

Sidebar: It took many years for agronomists to figure out and slowly pass on the news that wild hay actually reaches its peak protein potential in mid-to-late-August.

In 1947, my main contribution to the Hay Wars was to carry lunch to the men at noon. I had over a mile to walk and I didn't mind it a bit. As it was a hay day, the weather was always pleasant and the last half mile passed through Bill Rowland's quarter and through his yard. The trail was wide enough for a wagon and hayrack and had no wet spots. There was bush on both sides so there was always the warm smell of poplar, willows and hazelnuts, and birdsongs to keep me company. It was almost churchlike and I walked silently. There was no need to whistle or to pretend to be brave – no ravenous wolf would ever dare to disturb this solitude – and just when I least suspected it, I was attacked!

A little white-tailed deer came out of the bush and walked toward me. The locals called them jumpers and I had seen them from time-to-time while fetching cattle. Usually I only saw part of them – their white flag as they bounded away, but this little guy showed no fear. He stopped a few feet away and studied me with those soft, brown eyes. I put out a hand, he shied off a little, so I continued on with the lunch bag (a gunny sack) over my shoulder. In a few minutes I heard him following me with sniffs and little snorts. Soon he was at my side nudging my elbow and jacket pocket with his nose and stubby antlers. I was getting nervous. He wasn't hurting me, and his antlers were only fuzz-covered prongs, but I had never been this up close and personal with any wild animal, and I couldn't figure this one out. I started to trot and he soon gave up and turned back.

When I got to the hayfield and told them about this strange beast, they all had a good laugh. They told me that he had been orphaned last year and Mr. Rowland had bottle-fed him until he could eat on his own. When he had been nudging me he was looking for treats. Thereafter I always carried a sugar cube or a cookie and he often met me on the trail.

The poor little bugger didn't stand a chance. Most families (not ours) supplemented their meat diet with jumpers and Bambi was gone before the snow fell.

Occasionally Judy would accompany me on the hayfield lunch trek and one day when we got back to the home pasture, we stopped to build a little house in a poplar bluff. We kids always built houses – a snowbank house, or, on rainy days and winter evenings we would build walls of reversed dining room chairs and cover our house with sheets and blankets. I think now, that it was our way of creating a safe place to escape the often harsh reality of parental fights and arguments.

Anyway – this project was not a log house, but it was pretty neat. I always carried my jackknife on non-school days, so we picked a proper spot with small poplars at each corner. First I cut a few hefty willows and we tied them with binder twine between the poplars – four at shoulder height and four at the base. Then we started interweaving smaller and smaller willow branches until we had a nice, leafy, comfy enclosure to sit in. We worked quickly and steadily because chores awaited us at the house and we didn't want to catch heck. We didn't sit inside it for long. We would return tomorrow – we had plans for a willow roof.

Handy hint: Don't build your summer cottage in a cow pasture. When we came back the next day it was trash. Even the shoulder-high cross-pieces were gone – they were just the right height for back scratchers.

One warm sunny day in early June 1949, I ran away from home. Although, by my early teens I would be dealing with classic symptoms of manic depression, at nine years old I was simply pissed off. It wasn't school related – school had always been a haven – but life at home was getting more miserable every year. The Mom-and-Dad Wars were escalating and Rod, stuck on a farm he hated, was lashing out. I was caught in cross-fires left and right. Walter and Beth, always my protective rocks, were away working, so I decided it was time to make my own way in life, far away, and Winnipeg was the farthest away I could imagine.

Crafty me had planned it out well beforehand. I had twenty-five cents hoarded in my piggy bank – surely enough to last until my first payday. I was going to work at Hambley Hatcheries – they had all been so kind last year and I figured the oriental guy would show me what to look for on upside-down chicks. My backup plan may have been testing bikes at Ashdown's, but I really don't think I looked that far ahead.

D-Day arrived – a Monday school day. I always started the week with a clean shirt and overalls, and I needed no jacket in June. I had picked a school day because it would give me an eight-hour head start and in summertime we often didn't come home for lunch. Today I would leave before Judy and she could carry my sandwich to school.

After I passed the school-yard I was very careful to keep to the bush. I didn't want to meet any Richauds or Lowerys, and while I didn't have a wristwatch, I didn't need one. My inner sundial told me the time.

When I hit the highway I knew the other kids were at school, but I still had to be careful. I was going to hitchhike and if a familiar vehicle came along I was prepared to pretend I was walking west towards home. The first car was a stranger, so I stuck out my thumb and he stopped! This was easy, but I was not home-free yet. I held my breath until we passed the Red-and-White store in Deerhorn. Had we stopped there, my cover would have been blown to smithereens.

Phew! He didn't stop, and now we checked each other out. He was a travelling sales rep returning to Winnipeg. I was smart enough to ask him to drop me at Lundar – I knew short rides from town-to-town were necessary and I would have to make up a story for each ride. I told this guy I was going to visit my sick grandmother – yeah, right – Little Red Riding Hood in bib overalls.

At Lundar I took a break and walked down Main Street, looking in shop windows and avoiding eye contact with adults. I wasn't worried that I would be recognized – I was not exactly a man-about-town. I only hoped I wouldn't run into a truant officer. I'd heard about them, but I guess he was on days off. I did get the odd quizzical look and I picked up my pace – I was late for school again. I also avoided Mr. Balbon's welding shop – that would be a dangerous fellow-farmer infested place. It was not noon yet, but I had coins burning hole in my pocket, so I stopped at the Cash and Carry. This was not a brave move on my part – I knew that a man with cash in his pocket would never be turned away, so I bought a pop and a chocolate bar – five cents each. Then I spied a Captain Marvel comic book and made am impulse purchase – another ten cents down the tubes.

I sat on the wide steps of the store, chowed down on the bar, drank my pop and faced reality. I had a nickel left in my pocket, I knew payday would never happen, I knew Winnipeg was too far away, and I also knew the city held no answers to my problems. I was not sad at all. I had made my point to myself alone and I would return home and no one would be the wiser.

I decided to walk back on the CNR line. It would be a seven-mile hike to our side road but walking I was used to. I estimated I would reach school just around four o'clock and would go home as if I had put in a hard day hitting the books.

The walk home was great once I learned how to track-walk. The darned ties were not properly spaced. If I stepped on each tie my stride was too short, and using every second tie was too long. I could walk the rail but it was hard to keep my balance while reading Captain Marvel, so I walked beside the tracks. This was better, and I read as I walked, switching sides from time-to time because the grades outside the tie ends were sloped and my ankles needed an occasional change.

The rail line was far enough away from highway and houses to allow me to walk in solitude. The day grew warmer, a little hot even, and I could smell creosote, coal and coal-burning locomotives – nice, unfamiliar aromas. At Deerhorn the line passed near the Red and White – no action there this afternoon – and we now paralleled # 6. A half-mile later I turned off, crossed to our side road and reached the school just as kids were coming out. I was asked where I'd been, shrugged my shoulders and went home with Judy, who asked me questions I did not want to answer. I <u>did</u> tell her to keep her lip zipped, which she didn't, and which I didn't mind anyhow, because what was the point of running away if no one knew but me?

Believe it or not, I was never punished, but Mom and Dad were strangely silent when I was around. I had ditched Captain Marvel, regretfully, before I got home. Comic books were <u>not</u> allowed and I would surely have been whacked for that.

In 1948, we got a car – a beautiful blue 1947 Hudson Super Six four door sedan. Dad's Uncle Rob had passed away in Saskatchewan, Dad went out to the funeral and returned with the Hudson. Rob had written Dad asking him to come for a visit, but the letter took 3 weeks to reach us and Unc had croaked before Dad got there. I guess the old boy was miffed, because he changed his will before he departed. He was, as were almost all Saskatchewan farmers, an ardent socialist, and he left 30 grand to the Sask Government, no doubt receiving a penthouse reward in the CCF Heavenly Hilton.

That 30 thousand dollars would have been a huge game-changer for us, but the Hudson did its job – appearances count.

(CCF – Confederation of Canadian Farmers – forerunner to the New Democratic Party (NDP)

Sidebar: In late 1989 my wife and I dropped in at a CIBC branch to promote a loan to buy a C-store. At the time, we were driving our faithful ten-year-old Pontiac – still fully functional and fully paid for. The manager was unimpressed and turned us down, so with nothing else to spend our money on we traded the old girl in on a new GMC Safari – a dollar down and a dollar a week.

A year later we visited the same bank again, making sure the Safari was parked in front of his office window. It was not ours yet. GMAC owned everything but the steering wheel – but we got the loan.

The Hudson had two faults, one being that it hugged the road. All of the locals, from far and wide, drove buggy-springed Fords, and they handled mud roads and bush trails like ours very well. No matter how carefully the Hudson was driven on our turkey-track she snagged a couple of boulders but she was built like a tank and motored on.

The other fault mattered only to me. It had no radio – rats and double rats!

The biggest Hudson fan was Mom. She could grocery shop in style, which she did more often now, with Dad, Walter or Beth driving. She had seldom shopped herself pre-Hudson, and that frustrated her. Dad never got everything on the list and half the items were the wrong things anyway. The smooth-riding Hudson smoothed her life and we little ones got trickle-down benefits.

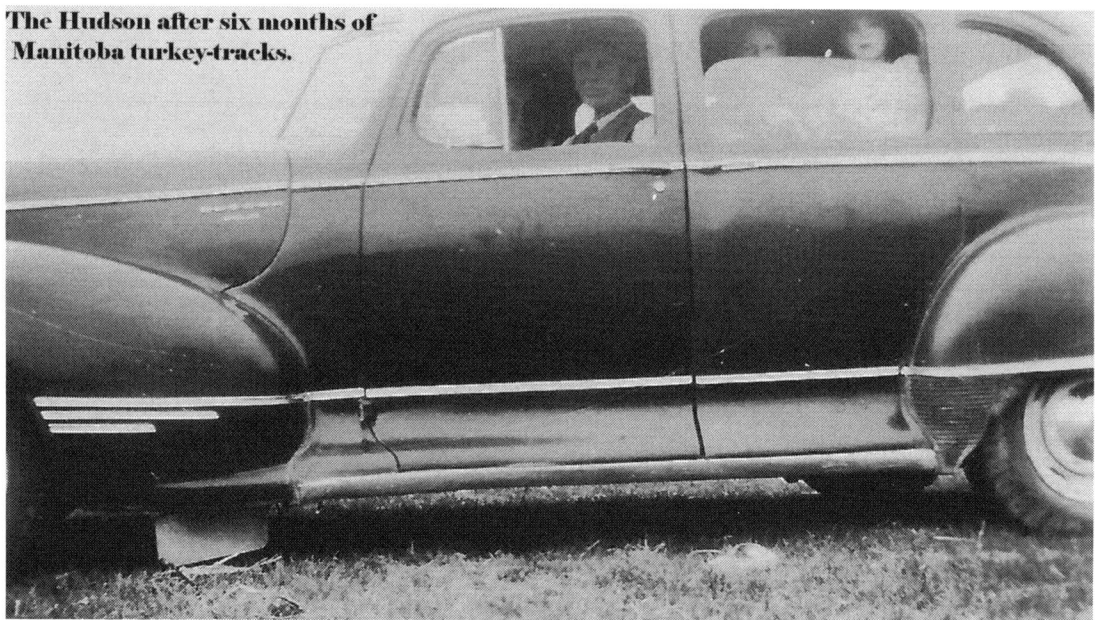

The Hudson after six months of Manitoba turkey-tracks.

Pre-Hudson, Mom used the Home Shopping Channel. Every year around late March or early April, if our road was driveable, the Watkins and Raleigh men paid Mom a visit (never on the same day) and one year one of them came on Easter break and I got to watch how it worked.

A large suitcase sort of thing was brought in, set on its end on the kitchen table and when opened, it had all the items on display on little shelves. Now Mom sat down to make her selections and I watched and hoped.

The first items held little interest for me. Mom would pick out baking stuff – little tins of paprika or thyme and little bottles of vanilla and lemon extract, for instance. Then came the home health remedies – aspirin, (ok) vitamin C capsules, (ok), followed by the dreaded Cod Liver Oil! (yuck) It always came in little, evil-looking bottles with a rubber bulb screw-on top with a dropper/pipette therein. Every morning Frank, Judy and I lined up, closed our eyes, stuck out our tongues and braced ourselves for two drops – only two, and one at a time. Mom made darned sure we swallowed each drop separately. It was pure torture, and happy were we when the cod liver oil ran out.

Always last, but not least as far as I was concerned, was the one-quart wide-mouth bottle of drink powder, either lemon or orange flavour. This was before Kool-Aid times and one bottle went a long way. Two heaping tablespoons per two-quart mason jar with sugar added, equalled many months of delicious hot-day treats and hayfield thirst quenchers. When the shopping came down to orange drink time I always held my breath, hoping the money had not run out.

And sometimes it did run out. The Captain controlled the vault and the salesman might show up any time – we had no phone for advance warning. If Dad was around, he would pull out his wallet, otherwise Mom had only her little cash hoard and it was never very healthy.

So now the salesman would close the display and go back to his car to bring in items to fill the order. One by one they sere itemized with a running total until Mom said, "Whoa.". Some items would have to be deleted – but not the cod liver oil – never! I'm sure Mom would have single-handedly slaughtered a pig to keep cod liver oil on the list.

And I remember so well that one year we had no drink powder.

(Frank's memory) One year, at the end of the list, mom said, "Oh yes, and a can of lemon pie filling."

The salesman countered, "I don't have any."

I started to cry. The thought of a whole year without lemon pies was just too much for a four-year-old.

"Oh, wait, I think I have one in the trunk,' he said, and went out for it.

Calamity diverted! He had likely promised it to another customer, but a child's tears are the hardest sell of all.

We had semi-strict dietary rules in force, based on health (primarily) and budget. There was always butter on the table, often home-churned, and store-bought bread occasionally replaced home-made. Sliced bread hit the shelves about then, but it cost five cents extra so I think the only loaf we saw was brought home for show and tell.

We had our own beef and pork but no power and no ice. Fresh meat left the table by mid-April. I don't recall smoking bacon or ham ourselves – perhaps a neighbour did so, keeping some pork for payment. We always had salt pork in summer and on Sundays, one of last year's laying hens got the axe. There was also home-canned beans and peas until the new crop came in. Along with these we had carrots, turnips and potatoes in the dug-out basement – (as many as had survived the annual cow raid, that is.)

Table Scrap: Beef Rings

In 1985 my wife and I visited Aunt Bessie, our dad's half-sister. She was in her eighties and lived in a "granny cottage" on her son's farm south of Brandon Mb. I had heard of beef rings and she explained the deal.

In the 30's and 40's beef rings were prevalent in rural areas such as in the Pembina Hills where and Bessie and Uncle Bertie lived at that time. Country roads were being steadily improved but electricity infrastructure was scarce.

Each ring was made up of six to eight families and every seven to ten days one family would butcher a beef. The meat was shared and it was an exercise in diplomacy. If you got a better cut, say steaks or a roast, you received a lesser cut or hamburger from the next kill. Bessie said it was a good arrangement and of course, rings only operated in the summer months.

(Beef rings would never fly in our Deerhorn community. If the roads turned to mud the meat would spoil before members could pick up their allotment.)

We didn't stay long that evening – Aunt Bessie was a bit tuckered out. She had spent the day "rogueing' flax with her son's family. Rogueing? We had never heard of it, so we got another lesson.

It had been too wet to spray weed killer and mustard had gotten a foothold in their flax field. They had spent two days walking the field, pulling mustard plants. Bessie, more than eighty years old, was still a "rogue".

Our breakfasts were always porridge – oatmeal mostly, with Red River Cereal or (ugh) Cream of Wheat with fresh un-homogenized milk and white sugar, and be careful you didn't try to cheat on the sugar or your knuckles got a rap.

Sunday morning (only Sunday) we got a treat – cereal and brown sugar. We had three main choices, Kellog's Corn Flakes, Quaker Puffed Wheat, and once in a while, Quaker Puffed Rice – shot from guns "kapow! shwing!" (No kidding) – Google will confirm this.

The cereal was always bought in Price Depot 45 gallon drums (slight exaggeration.) Even though this was pre-chemical times, they seldom hatched creepy-crawlies before the last of the large bags were devoured.

Fresh fruit was always available but very, very strictly enforced – one apple or one orange a day. Dad brought them home in wooden crates which were carried down the ladder to the dug-out cold room and the daily allotment was retrieved every day after supper. Scam artist me would gladly lift the trap door and climb down to fetch them. It was dark down there. "What's taking you so long?" someone would holler. I never answered, because I was chewing a large bite of apple. Then I would stash it in a safe place for the next night – an apple and a quarter a day kept my doctor away. Those boxes lasted well over a month and never did more than one or two apples or oranges turn bad. The last Macintosh was just as crispy and fresh as the first. Now they barely survive the display cooler – that's progress?

In 1950 just after school was out for the summer I got a job! Eleven years old, and I had a job – what excitement!

Actually Dad got me the job. He was on the municipal council and they were doing some improvements on some rural roads. The preliminary work was dirt movement to bring certain sections up to grade and it was being done by cat and scraper with no road grader back-up. The "scraper" had a big hydraulic actuated scoop and the cat skinner scooped up dirt and spread it on the road, killing two birds with one stone -- ditch and grade. My mission was to follow the scraper and toss any rocks larger than a hardball back into the ditch. I would earn fifty cents an hour to play ball – good deal.

The first job was on the LaFontaine Line, approximately four miles east of Deerhorn. Dad drove me over each morning and picked me up after my eight-hour day. The cat skinner was a guy from Lundar and he didn't seem to like me very much. He never smiled and never said a word to me. Once or twice, as he scooped a load in the ditch beside me, he pointed to a rock I had missed – which I hadn't – I just hadn't got to it yet. We ate our lunches alone – he sat on the cat and I sat on the shoulder of the road 50 feet away. This didn't bother me any – I thought it was how grown-ups did these things.

We worked two days on the LaFontaine Line and on Friday evening I was tired but satisfied. I already had eight dollars coming and I had never seen that much money in my whole life!

On Monday we moved to our road. The whole road needed work but this year the municipal budget could only handle a two-hundred-yard stretch from the edge of the slough near #6 westward.

Dad didn't have to deliver me now. I rode the CCM to work and stashed it in a clump of willows well off the road (I thought) and because I could see the stakes marking the west work limit, I was sure the bike was safe.

It was not safe! On his last pass before quitting time that cat skinner made his turn-around much further west than necessary. I was tossing my last rocks of the day and I hoped – really hoped that the willows he had trampled did not hold my bike.

When he came back past me to dump the load he smiled at me – the first time he had done so and it was a mean smile and I knew my goose was cooked. I ran back to find a mangled CCM double-bar that had been in the family for years. It didn't even look like a bicycle now. The cat treads had run directly over it and it would never carry anyone again – never! I carried it home, bawling all the way. At home I took the bike to an empty granary, sat beside it and continued crying until I was out of tears and my tummy hurt. I wouldn't go in for supper. I wasn't hungry and I knew I would be told it was my own darned fault. When Walter came home he came out and sat with me. He said they would buy me a new bike, but somehow I knew that would never happen.

Of course, my rock-tossing career was over. I couldn't stand another jeering grin from that peckerhead cat skinner and to ice the cake and to add insult to injury, when I got paid I never saw the money. Dad had a twelve-dollar check for me which I had to endorse so he could cash it for me but he kept the cash. I learned two things at my first job. Never trust anyone – even your own father.

Chapter IV
Fiddle Foot Training – 3rd Semester

But hey – we were moving again! Because I was eleven now, I picked up the vibes a bit sooner than the last time. A search had been underway for a few months and the decision was made. We were moving to a dairy farm in Northwestern Ontario. The farm was five miles west of Fort Frances – a paper mill town of 7,000 souls on the border of Northern Minnesota. We would be shipping milk to a dairy in this virtual metropolis! It would be my parents last move – and it was a good one.

We moved during the Christmas school break in December, 1950. The two-ton once again hauled stuff – mostly furniture and cows. Other possessions and some cattle were shipped by rail in a "farmer's car".

(Farmers' cars were unique and would soon be phased out. They were a combination dealie, one end was like a boxcar and held furniture and the like. The other end was a cattle car – neat, huh?)

And so, we drove back into the Canadian Shield in our 1947 Hudson Super Six, on past Kenora to turn south on Hwy 71 – that road which we had passed by four years before without a second glance.

Sidebar: At the junction where the north end of 71 hits the Trans-Canada there was a huge stone with a bronze plaque proclaiming it "The Great River Road," the theory being that by travelling 120 miles south to Fort Frances, crossing the bridge to International Falls Minnesota, and thence onward to Bemidji or Minneapolis/St. Paul, one could then wend one's way south following the Mississippi – sort of far-fetched, I always thought.

We drove south on 71, twisting through typically convoluted Shield country. The Ice Age glaciers must have missed this stretch. There were lots of hills, lots of curves and many, many frozen lakes. On down past the west side of Lake of the Woods we drove, through Sioux Narrows and Nestor Falls and twenty miles south of Nestor we rounded a curve in a rock cut – and just like it had been cut with an axe, we were out of the Sheild and into the Rainy River valley. Ever since and every time I've driven that road I am still amazed at the sudden change.

Thirty-five miles later we were at our new (old) farm house – another two-storey deal, but it was of conventional wood frame construction, and even though it was heated by wood it had a real furnace in a real basement with real heat ducts.

This would be my home base for many years and it was in the best of all possible worlds – good farms, all with good wood lots, and a paper mill on each side of the Rainy River spewing out steam, smoke, rolls of paper and paycheques. The main highway was paved, sideroads were well gravelled, and in town TWO movie theatres ran two shows every night (except Sunday.)

The farm was in the municipality of Alberton but no one called it Alberton in 1950. It was Crozier in those days, and Crozier was not a town, village or a collection of houses. Crozier was Hampton's store and Mr. and Mrs. P. F. Hampton had the post office and the post office address was Crozier because that was the name of the recently closed railway station a half-mile south of the store. It was Crozier to us until we sold the farm in 1976, but if you ask a young (50-year-old!) person now if they live in Crozier, you'll get a blank look – It's Alberton these days.

It must have been a hectic move, what with well-below freezing temps, Christmas coming and such. I don't recall the two weeks very well and Christmas must have been a minor celebration but it didn't matter. What was important was that we had a proper barn with a full hay loft and a semi-proper house.

The house actually had proper exterior siding, which was somewhat weathered and would be replaced within a few years. The first floor was finished with a huge kitchen/dining room, a living room for our piano and stuff and a bedroom for Mom and Dad. Upstairs was an as yet unfinished, uninsulated partition-less ice box with boards laid un-nailed on the floor joists. As in Deerhorn, we slept two to a bed under layers of blankets and quilts. This was no biggie to us – we knew the drill.

The barn, as it should be, was new and shiny. On any dairy farm the barn should be new and shiny – the barn makes the money. This one was as big as a barn with proper stanchions for twenty-two cows, a two-horse stall, two equal sized calf pens, and a 6'x8' area for the cream separator and vacuum pump. (We had a milking machine, by gum!) Also, ladder rungs nailed to wall studs led up to a hayloft full of sweet-smelling alfalfa and timothy hay. This barn was so clean and comfy I could have lived in it.

Mom on the path to our shiny new barn

Our well was like lemon icing on a two-tiered cake. This was the well of our dreams, an 8" steel crib baby drilled half-way to China. On the plank platform and extending into this fountain of life was a big hand pump with a three-inch spout! No longer did cows have to be watered in shifts. No more carrying pails of precious hot water from the kitchen to pour down the pipe to thaw the frozen cylinder. This well never froze!

And it didn't take long for the bloom of my rose vis-a-vis the well and pump to fade. On non-school days it was my job to man the pump handle and I never could understand how twenty-two cows and two horses could drink that much water without peeing their pants.

We started shipping milk lickety-split. The previous owner's husband had died just a few months before and she and her teen-age daughter had been doing their best to keep the four cans per day contract filled. They were almost tuckered out – he had been sick for a long time. In fact, the hayloft had been filled last summer in a matter of days with a good old-fashioned neighbourhood "haying bee."

I had been helping milk cows back in Deerhorn, mainly by carrying pails of milk back to the house where the cream separator sat in an enclosed porch beside the kitchen door. I was also the main muscle on the separator and it was a relentless task-master. On the base of the handle was a bicycle type bell and a tube on the handle shaft held a steel ball. If you turned the handle too slowly the ball rang the bell, too fast and the ball stayed at the far end. The trick was to keep the ball in the middle and that was hard for a kid to do. But now, the milk contract came first, cream was separated only when we had surplus milk and I was promoted to milk-man apprentice.

At first our herd was a mixture of medium to good producers. Our best cows had been brought on another two-ton trip or two and replacement heifers came on the CNR farmers' car. Rod stayed at Deerhorn until the last of the beef cattle were shipped to Winnipeg stockyards, caught a ride in on the cattle buyer's last truck and I'm sure he didn't wave goodbye to the now lifeless Deerhorn buildings.

Sidebar: In 1950 the Manitoba Department of Agriculture had completed an aggressive program to rid cattle herds of TB infected animals. Cows, calves and bulls that tested positive were shipped to the stockyards immediately. (Bovine tuberculosis does not compromise meat quality.)

This purge flooded the beef market and prices were rock bottom, alleviated slightly by government subsidies. We were pretty lucky in only losing two or three head. We were also fortunate in that the tests were done before we moved. (Ontario was TB free, and without the paperwork to certify our herd, the tests would have been very time consuming, making the move un-doable.)

Our Deerhorn neighbour Joe Rivet took a body blow – his milk herd tested 100% positive. Joe had converted his one-time chicken house to cows and what he didn't know, nor did anyone else, was that chickens can carry a mild strain of tuberculosis. The hens could live with it in their blood but the cattle picked it up. Truth be told, it probably didn't bother the cattle much either, but no matter how mild the reaction the milk must be totally TB-free. The only part of the herd that Joe could keep were those that had been housed separately.

Getting into the proper milking routine was quite a kerfuffle. To really understand what we were up against you have to imagine an expansion pro football team in pre-season practice.

First we had the Executive Branch. Dad wore many hats – President, Chief Financial Officer, General Manager in Charge of Player Acquisition/Development and Head Coach.

Walter was Vice Everything, Assistant Coach and in charge of Player Movement, Rod was versatile, and filled any position as necessary. I was Water Boy and Frank was the Grade One rookie prospect. (And on our team human females were not allowed to participate.)

The team was a collection of veterans – some had come with the farm, some had come from Deerhorn and others were signed on as free agents. They all had to learn the playbook, get to know one another and the Executive Branch.

The Crozier cows, on their home turf, knew where they stood in the barn, knew where to find the well and water trough, and knew the milking machine – a pre-war Massey-Harris double unit. The new-to-the-barn cows knew nothing and now the fun began.

First of all, the new arrivals had to be introduced to the stanchion concept. (At Deerhorn, they had stood at the manger, tethered by neck chains.) Stanchion training was not too difficult. When they came back from the water trough they were herded into their slots and after a day or two they had that figured out with help from the Crozier herd. If a new-comer tried to expropriate some real estate, she got a head butt in the belly. It also helped when they learned that on their return, tasty morsels of oat/barley chop awaited.

(One cool thing I got a kick out of – there was always a wily entrepreneur in the herd. She made sure she was first in line when the barn door opened and she would steal a bite from a spot or two before going to her own slot. "Who, me? – A thief?" she'd say with a grin.)

But the "know your own spot" was a bit of a problem. The Massey milked two cows at a time. Those cows that were used to a milking machine stood patiently as the teat cups were attached, but the Deerhorn cows didn't like it a bit. They kicked – and they didn't care if they caught machine or man. If you got one cup attached, it would often be kicked off before the others could be put on.

The answer was to pair an experienced pro with a rookie – easier said than done. Now two or three cows had to learn new spots and they were reluctant to do so, but by the end of January things settled down.

(Dad didn't fully trust the Massey, and rightfully so. It was not designed to drop off the teats when all the milk had been extracted and if careful attention was lacking the cow could be injured. We had strict instructions to remove the suction cups early and "strip" the cow by hand. Five years later a modern single unit Surge milker was added to the equipment list and it never over-milked a cow.)

One of the Deerhorn cows – Old Humpey – never did accept machine milking, but she was a good producer and always dropped a fine calf. Dad cut her some slack and milked her by hand until she was finally retired (at a much older age than usual.) She also carried the cowbell. Every team needs a quarterback and stubborn Humpey did her job well. She led the herd to the water trough, got first dibs and led the way back. In summer she would be the first to recognize the open gate when pastures were rotated. In the fall after the hay was brought in, the cattle ranged further away and when Humpey saw me coming she called an audible and started the herd home well before Mickey and I reached them.

When the old girl went to Winnipeg to be converted to bologna I missed her. A pretender to the throne was awarded the bell-crown, but was never up to Humpey standards.

One great disappointment to me was that once again we were powerless. It came as no surprise – it was one of my first questions last fall, but nevertheless, I was miffed. The house was wired for electricity and useless light switches, outlets and ceiling bulbs reminded me every day, and every day I imagined a day in the future when a switch would be flicked and a light would come on. I wanted to know why we had no juice and the explanation was a familiar story. Ontario Hydro, as in Manitoba, would build only the first half-mile of line. If an already serviced house was in that first stretch then the half-mile rule kicked in again, but we had no neighbours between us and the highway three quarters of a mile away and thus would have to pay for the last quarter-mile. The farm finances were still recovering, and the cash would not be found until three or four years later.

I shouldn't have complained – Mom must have been very disappointed. She still had to cook on a wood stove, wash clothes by hand and iron with little iron boats. The only advantage to her that I could think of was that now she could shop regularly and keep some stuff cool in a proper basement. Small potatoes for her.

The happiest new Crozierite was Rod. He could now continue his interrupted high school education, which he did in September of '51.

We young'uns started at our new school on January 2nd. I was in grade six, Judy in grade four and pain-in-the butt Frank was in grade one. He was actually a pretty good kid, but we now had over a mile to walk to school and he was pokey. I got my orders from Mrs. Captain – never leave the troops behind.

The school was a typical one-roomer of that era, but I was impressed. It was much larger than our last school and sat high on an actual cement basement. Wide steps led up to an 8 x 12 roofed verandah and inside to the left of the door was a small doorless room with hooks for boys to hang their parkas. To the right, stairs led down one flight to a side door and switched back to the basement. Upstairs, we entered the classroom through a second door where the girls turned left to their own "cloakroom" – their private lair with a door that closed behind them. This was a sacred place – off limits to us guys and I don't recall ever getting more than a glimpse of that sanctuary. I could never understand why it was a "cloakroom" – only Sherlock Holmes wore a cloak.

There were five or six rows of single desks and the desks in each row were attached to long 1" x 4"s - sort of a "super-sled." I suppose this was so the whole row could be moved to scrub the floor.

The west wall was almost total glass (it seemed) and like Deerhorn they were single-paned. I have a faint memory that storm windows may have been used in winters but were never replaced with summer screens. With school years ending in June and starting in September, only our imaginations got overheated.

Both north and east walls were windowless. There was no room for windows – It was all blackboard. With attendance hovering around the 25 to 30 mark the teacher needed all the blackboards and I'm sure the Crozier (Alberton) #3 chalk and eraser expenses equalled Deerhorn's total school budget. Every day, once a day, the suck-up of the day got to take the erasers outside and bang them against each other to clean them. You always faced downwind when you did this, otherwise you'd return looking like Frosty the Snowman.

The teacher's desk, as usual, was front and centre beneath the King's portrait. In the northwest corner sat a piano, always more-or-less in tune, and Miss Wheeler would hammer out a pretty fair accompaniment as we sang Oh Canada and God Save the King, which he didn't, because within a year or less the portrait was changed.

The basement held a stove/furnace and in each north corner were biffies – boy's and girls. Although the school had lights – fluorescent fixtures upstairs and ceiling bulbs downstairs, there was no plumbing. The biffies were basic one-holers, rarely used by the students and I can only assume that before the concrete had been poured a very deep hole had been dug.

Another thing – the school had central heating (actually, slightly off-centre.) The stove in the basement must have been a big one. It took four-foot hunks of birch, and the outside woodpile was cured birch only. We could only see the door on this monster, as the rest was surrounded by zinc-anodized tin which continued up to the ceiling where a cast-iron 4' x4' grate sat just inside the classroom door. A chain led from the stove's draught control to the wall upstairs and the teacher controlled the draught control.

Here's the drill: Henry Miller, who was in grade eight and who lived a ¼ mile to the west, was the fire-starter. Every winter school-day he would light the stove at 8 am, fill it with birch, set the draught at minimum and go home for breakfast. For this he received the princely sum of three dollars per month, and he may have continued these duties when he started high school.

Now Miss Wheeler arrived around eight-thirty and opened the draught. By the time we kids started straggling in, the school was warm and toasty. Most of the time an older boy would toss in three or four sticks at noon, but on bitter cold days a west wind could find its way through nooks and crannies and the stove needed more than one transfusion. This was not a forced-air deal, but the hot-air convection worked just as well. When we came in for recess or at noon-hour we would jostle for position over the grate. The uprush hot air would actually make our parkas bell outwards. None of us could stand the heat too long, and of course, it was women and children first.

Recess and noon-hour was always outside play-time for boys from grade four to eight, and only on the coldest, windiest days were we allowed to play in the basement. We didn't care – it was boring down there with "ugh" rugrats and "double-ugh" girls.

Outside we had a hockey rink to play in. The rink had been flooded and was ready for competition by mid-December but we were not allowed to bring skates to school. Most of the time we played pretend hockey and took cold, shivering breaks in the rink shack which had a stove which we were not allowed to use. Sometimes during the week, we would start to shovel snow off the ice for an always exciting Crozier vs always sub-standard neighbouring towns on the coming weekend. We soon powered out – older young men would finish the job later.

(Now – a word of caution. Anyone lucky enough to be reading the following must be prepared to believe anything, but believe me, it is absolutely true. It is a "right of passage" story – a symbolic journey to possible non-manhood.)

First some background is needed. The skating rink may have been regulation size, but the side and end-boards were definitely non-regulation. For some reason, perhaps to foil the winter winds, the boards were six feet high. Now some farm-bred ingenuity came into play.

Every farmer knows and knew that a flat-top post may soak up water and rot – especially if planks were nailed flat on the post-tops. To counteract this possibility, the rink posts had angled tops and a 2x6 was nailed on the angle.

So we would walk the 2x6's and as sure as God made little green apples one foot would go one way and one foot the other, resulting in one blood-curdling scream and much laughter from luckier boys. The victim rolled around in the snow hoping the bell didn't ring soon. After recess we never complained to the teacher even if we had difficulty standing straight and we never told the girls (of course) or the male basement rugrats either. Next winter they would get their own lessons. We all had at least two painful lessons before we got a handle on the "tightrope" concept but no permanent damage was done – we all grew up to be fathers.

Everybody walked to school back then – everybody. The only school bus we saw was an old blue-and white Western Flyer passenger coach that hauled kids to high school and it seldom left the pavement. We walked, and the exercise was good for us. (Nowadays, school buses stop at every rural driveway. Town kids walk farther than their country cousins.)

We usually walked together. At the highway we three would join nine coming south on Hammond road and on the next half-mile east on the highway we would pick up five others. From the east came another 13. In the mornings we were spread out like wandering nomads but at 4pm we hit the highway shoulder en masse. We were told to always walk on the side facing oncoming traffic and to stay off the pavement, which we usually did if we felt like it. However, we were smart enough to stick to one side or the other.

Highway traffic was lighter in the '50s, but it varied seasonally. During winter months pulp trucks streamed by, most of them single-axles carrying up to five cords. The wood haul slowed down somewhat by the end of April but when the lakes thawed at Nestor Falls and Sioux Narrows big American cars picked up the slack. Adults never admitted to us that they were concerned, thus we never thought of the possible danger. Common sense kept us safe. There was no way anything could have been done, anyway.

– you can't post a twenty-five mile per hour limit on two miles of highway. I suppose signs could have been erected – "Danger – Watch for Semi-Wild Animals." The only similar signs I had seen were in rock cuts on Hwy 71 – "Danger – Watch for Falling Rocks." I guess rocks were more important than kids, (and although I always watched, I never saw a falling rock.)

I sort of enjoyed the walk to school, especially the highway traffic. Cars and trucks streamed by us and after four years of Interlake isolation this was great! I was already a neophyte motorized transportation buff and now, with a lot of help from my friends, I was brought up to speed.

At that time, every truck from pickups to highway haulers had to have the owner's name on its doors and that helped. Chevrolets, GMCs, Fords, Mercurys, and the occasional International or Dodge went back and forth and we knew them all and knew them personally (we felt) – names like Paldio, Gerula, Wardman and Lougheed. There were also a few tractor-trailers in the mix. C.V. Strachan and Sons had two L180 Internationals, Ivan Jewett a Mercury Big Job powered by a huge Lincoln flathead, and Emo Forest Products a Ford-powered unit. One unusual truck on any Canadian road was Jim Witherspoon's White "Twin-Power" and though it was also red, I was well past that confusion.

We knew them all and we knew everything about the owners. (We did not.) We knew who hauled the biggest loads and the most loads. We thought we knew who was making money and who wasn't by how new their trucks were. What we didn't know was that while they were all making a profit, the owners of the older trucks were keeping their cash.

In the winter months the car/truck ratio was about 50/50 and we had a handle on half the cars, bringing our overall average down to 75%, still good enough to hit the honour roll.

And then spring arrived, the lake ice turned to blue water and the highway turned to shining chrome. The school sat on a rise and as we walked from the west a continuous ribbon of flashing grilles poured over the ridge ahead of us. Most of them were big cars, many had whitewall tires, full wheel disks and radio aerials, and all of them were new. They carried different licence plates – Minnesota (Land of 10,000 Lakes,) Wisconsin (America's Dairyland,) – and Ohio – and Iowa – and more. I thought every rich person south of the border was coming north this year and the older boys laughed – this had been old hat for them since the end of the war.

Kids are naturally competitive so we compared eyesight and recognition skills as we walked. When an oncoming car's front end popped over the hill we would call out the make, model and year and we were seldom wrong. I wonder – did the other boys dream of someday owning one of these land yachts? I sure did.

We walked that highway every fall, winter and spring. If it rained, we toughed it out – few of us had raincoats. If it was pouring cats and dogs we would hang out on the school porch waiting for a break, and then we'd hightail it. Sometimes a parent would show up, but they always came from Hammond Road. There wasn't enough room for all the northwest bunch, so their own kids got a ride with the littlest neighbours stuffed in –

the rest of us could sink or swim – and swim we often did. One rainy day Judy, Frank and I were trotting along with the Hill kids. Mr. Hill came by on his way home from the mill, picked us all up and was good enough to go out of his way to drop us at our door – what a kind man.

The winter walk home facing a cold west wind could be brutal. We held our parka hoods close with our heads down, and on those days we walked single file with the roadside snowbank near at hand. If a wood truck came by, adding a bonus gust of wind, we stopped and turned our faces toward the ditch. We Durnins were always glad to turn south on our side road – soon the bush would protect us.

So we walked, we never got hit and we never got into trouble – except – one day I did.

I was in grade eight and it was coming spring. A warm spell in late March made for balmy days and melting snow. Now we walked with our parkas open, throwing snowballs at each other – which could be dangerous because if any kid should duck and stagger onto the pavement – well – let's just say we didn't want that to happen. So we picked on bigger targets – wood trucks. At first we were content to fire at the loads but things naturally escalated and we pasted a hood or a door, knowing full well that a loaded truck could not stop in time for the driver to run us down, nor would he turn around to chase us. It was great fun until my brain cramped.

A car came along - an old car, a slow car, and I knew the car. It was owned by a family from the next village to the west and we snobs knew they were poor folks – "on welfare," our parents would say. As they came by I fired one and my trajectory deflection was right on. I hit the side of the car's hood!

Holy cow! He stopped – snubbed her up on the spot – that car had good brakes! Before he opened his door Hammonds and Hills, and Degagnes and Durnins, big and little, jumped the guard-post wires and tumbled down through the snow towards Kitchen Creek below, and I stood there, because if I ran I could not hide. I wouldn't run anyway, because whatever was coming, I deserved it.

He was irate, I'm telling you, and I got the works. He pointed to the open window on the passenger side where his wife sat holding the latest baby. What if I had hit the child, or his wife, or had missed them and hit him, causing him to lose control. Then he turned me around and gave me a swift kick in the butt. The kick hurt neither my bum nor my pride. I got laughed at but I'd taken it like a man. On the way home I threatened mayhem if Judy or Frank ratted me out because of they told Dad, I knew I was a dead man. The worst part of the whole deal was waiting two or three days until I was sure I was off the hook.

Over a span of almost forty years I would run into that gent from time-to-time. We always smiled and said "Hello" and I always added, "Sir." He had dignity that day at Kitchen Creek and for the first time in my life I understood what dignity meant. That ass-kicking taught me more than I ever learned in school.

Table Scrap: As Told to me by Bob Dimit in 1993

Bob and his wife Gwen drove school bus – each on different routes. Among others on Bob's bus there were always kids from a large family. One day one of those boys kicked up a ruckus and got kicked off the bus two miles from home. It started to rain before Bob reached the kid's house, so he stopped and went in to see Mom, telling her that maybe she should go pick the rascal up.

"I don't think so," she said, "I think a wet walk will teach him a lesson."

Two weeks later Bob's wife ejected a young fellow, an only child, a quarter mile from home on a nice day, and when she got home the phone was ringing. The school board wanted to know what was going on. The boy's mother wanted Gwen fired, and they would lawyer up and a lawsuit was mentioned.

You tell me – who was the better parent?

When May 1951 rolled around with ever-longer evenings, disappearing snow and a perfectly good non-muddy side road beckoning, I lobbied Dad and he bought me a bicycle. It was a pale limitation of the strong CCM, but I was looking for function, not form. Now, with wheels and good roads to ride on, my world started to expand. On nice evenings with supper over and the milking done I would peddle to the Miller's on Kitchen Creek. Sometimes we would play board games inside the house, but more often we would head to the baseball diamond at school. Other kids might show up and we could have a game of work-up if anyone had brought a ball and bat which I don't think anyone did because not one of us owned stuff like that. I think we relied on temporary petty theft. After afternoon recess we would "forget" to return the school board's ball and bat to the teacher. If that ruse failed, we'd use a stick and small stones and learn to duck. Friday evenings were the best time to go there. Then older young men would show up and we would hope to be invited to play. We were outfielders – far out – somebody had to chase down the home runs.

One Saturday I rode over to Miller's and Tommy (who was my age) introduced me to environmental recycling. We rode our bikes the half-mile to Hampton's Store picking pop bottles from the ditch along the way with Tommy on one side and me on the other. At the store Mrs. Hampton gave us two cents per bottle, which we immediately reinvested in pop, which we drank on the store steps so we wouldn't have to pay the deposit.

I was hooked. The following Saturday we went west one mile then returned to Hampton's. This time the pickings improved because no one ever walked that west stretch.

We made plans at school and our next foray was an ambitious trek – 3 miles west to Johnny Canuck's at the Lavallee River bridge. I was a bit on edge that day. I had failed to get permission from Mom and Dad.

For three summers Tommy and I policed the ditches and it was fun and profitable, because as neither of us got an allowance, it was our only spending money. We found that once we had finished spring clean-up a weekly trip was unworthy and when haying started I had to help, but it worked out that we averaged one or two scavenges per month, sometimes hitting both stores on the same day. Beer bottles were also picked, but I let Tommy take them home. His uncle would cash them in and actually give him the money.

We arranged sort of a "Buyers' Club." We both liked comics, but I couldn't take one home, so Tommy would buy one. We both liked bubble gum, so I would buy two packs of "Topps" with four baseball cards in each. I would share the gum with Tommy and he would let me read the comic book while he studied the cards. If we had a good haul we would each have a pop – if not, we would share one, usually an Orange or Grape Crush.

We never picked up litter because there was none. Canned soft drinks didn't exist then and potato chips were unheard of, as were plastic bags. The only thing we left behind were American beer bottles and if we shared a Cracker Jack the empty box was left at the store. For three years we "adopted" four miles of pavement.

In the summer of '53 I started smoking, thanks to my brother Walter, our Hudson and my budding felonistic tendencies. The Hudson was a smuggler's dream car come true. The hood lifted from the rear (Ford "introduced" this safety feature ten years later) and because the pivot point was 12 or 14 inches from the front, when the hood was lifted its nose ducked down behind the grille. The Hudson was a big car and a "straight eight"was available, but our Super Six engine was shorter and the radiator sat two feet behind the grille. Thus a dark mini-cavern could hold "duty-free" contraband – good quality cotton tee shirts, reasonably priced denim jeans and cartons of Pall-Mall cigarettes. (Walter smoked Pall-Malls.)

He bought them by the carton – ten packs of 20 cigs, and he always opened one end. I waited until a new carton was opened, and one Saturday morning I made my move. When I was sure the little pests were busy elsewhere I went to Walter's room, took one pack of smokes and by gently tapping the carton on it's open end it now looked un-tampered with. The empty space was at the back end.

Now what to do? I hadn't thought this out, so I stashed my loot and studied on the problem. I knew that Judy or Frank would eventually find my hiding spot and gladly wait to turn me in when they needed to plea-bargain their own way out of a future transgression. I could take a chance and stash it in the hayloft but I would never smoke up there – I <u>did</u> have some smarts, so I decided to share them.

Alan McGuire, a foster kid, was living with the Sparks in a house on Kitchen Creek (where the golf course clubhouse now sits.) He was my age, a good guy, and we got along well. It was a nice September Saturday afternoon and with no chores pressing I asked for and received permission to go over and play with Alan.

Now I had to figure out how to sneak the cigarette pack out of the house. If I put them in my shirt or jeans pocket the bulge would show, but because it was September I could take a jacket just in case the day cooled off so I tucked the Pall-Malls into a pocket. But all evil plans have a fatal flaw – I forgot matches.

At Alan's I told him I had smokes but no fire. He said there were wooden matches in Mrs. Sparks' kitchen, but she was an ever-vigilant type. We waited until she left the kitchen momentarily and Allen scoffed three matches before she returned – only three.

Now what? We could go to the little barn but it was too close to the house and Mrs. Sparks would surely spot the smoke signals. If we went across the road to share with Tommy there were always little Millers around and those loose lips would sink our ship. We decided to walk south on Kitchen Creek to find a cozy spot to puff and talk about the good old days.

We sat under a poplar on the east bank with the afternoon sun warming us and lit up. There was no wind and the poplar leaves dissipated the smoke. And now we had another problem – two matches and 18 cigarettes left. We could light up two more leaving us one match, but what if that one failed to strike? Plan B swung into place.

We chain smoked nine cigarettes each, lighting the next from the previous butt and by the time the last cigarette (which didn't taste so good after all) was done, we were dizzy and queasy in the tummy department. We said "So long" and went our separate ways. I didn't trust my rubber legs to jump the creek, so I walked south to the railroad tracks and used the bridge.

I started smoking that Saturday at 2 pm, quit smoking at 3 pm, and never touched tobacco again until eight years later.

Sidebar: That Hudson, although you couldn't tell by looking at it, was a huge car. A lot of cars of that era bragged on holding seven passengers, but few lived up to the hype. The Hudson, however, had ample room for three in the front and four in the broad rear seat.

It rode like a dream. Hudson had been using double control arms and coils at the front for years, and the control arms were long. Other brands, Including Ford (when they put the buggy springs on the shelf) used shorter arms, resulting in a rougher ride. The Hudson's rear leaf springs were also longer and the trunk was huge. When Rod went back to high school he drove the car every day. Busch's Transport had been picking our milk up every morning, but with Rod passing Flinders' Dairy on his way in, he now hauled the milk. Four eight-gallon milk cans fit into the trunk with the lid closing securely! If the dairy took a surplus can then the spare would be removed, leaving room for the fifth can! No other car at no time could match that.

A couple of years later before the Hudson was traded for a '50 Meteor, Dad bought another 2x8 gallon contract. The two extra cans now rode daily on the back seat floor, upright, and with little loss of front seat leg room – amazing!!

Chapter V

Other Vignettes from the 50s Follow – With appropriate headings.

Me, the Rifle and a Close Call on the Cattle Drive.
Our family armament was a long-barreled single-shot Winchester .22, used primarily to knock off a pig or steer when fall slaughtering time rolled around. I was occasionally allowed to carry it, usually in the fall when the cattle were further away. I think it was my job to drill a magpie or two, but they steered clear of anyone carrying anything that looked like a rifle. I wouldn't shoot songbirds or groundhogs and in fact, didn't shoot much of anything – bullets cost money, doncha know.

The back pasture gate was open now allowing the cattle to graze on hay stubble. They had two fields to choose from and I never knew which one held their flavour of the day. There was a little stretch of bush before the hayfields and in the bush near the fields sat an old "bucking pole" ramp which had not been used for many years. A lot of the boards were missing but I could climb the slope quite easily, and because it stood 12 feet high at the end it was my lookout point.

I saw that the cows were nearby and there was time to spare before Mickey and I took them home. I took a break, soaking up the late afternoon sun, listening to birds talking, and below me wild bees were wrapping up their own summer. It was so peaceful – and then I spotted a nail.

The board it had held was long gone and it stuck up about one inch – probably a 3 or 3½ incher. I wondered if a .22 would cut that nail, so I tried it out. I held the muzzle within two inches of the nail, pulled the trigger and an angry wasp buzzed past my left ear. Holy cow! – the lead slug had only bent the nail enough to air mail itself almost straight back, and I never tried that again.

Looking back, I have to think the bullet would have had to hit me in the eye to do permanent damage. Anywhere else would have only caused another ricochet – I was as dumb as a sack of rocks.

Prairie Flatlanders Cutting Pulpwood
We started cutting pulpwood in the winter of '51/'52', something no one in the family had ever done before.

The wood lot was at the west end of our half section. The hayfields sloped gently down to an ash swale, which became a cedar swamp which became part of a huge spruce swamp which extended three miles west and ran south almost six miles. Our slice totalled about a hundred acres and it would supply us with firewood, fence posts and non-dairy income for many years.

First things first – Dad needed a contract – and this was easy. Although the mill usually dealt in multi-cord contracts with larger outfits on crown lands, the wood buyer, Ed Langstaff, always gave small contracts to farmers – he was an ex-country lad himself. Also, because he knew we were new to the game, our contracts never exceeded 50 cords of spruce and 20 to 25 cords of balsam per winter season.

Next on the list was equipment – a necessary investment which did not require a bank loan or line of credit. A second "swede saw" and a second axe was purchased.

Horsepower was supplied in-house by Bess and Doll, a mother/daughter team we had brought from Manitoba.

The biggest investment was to modify our sleighs. At Deerhorn they had only carried the hay rack to move stacked hay from field to hay yard and the rear set was attached to the front by a simple "tongue" with a pivot point. This setup worked in the hayfield and on wide trails but the runners did not follow the front when cornering. This was unacceptable on twisty bush trails, so the tongue was replaced by cross-chains.

Manpower was also supplied in-house – Dad and Walter during the week with Rod and I added on weekends and over Christmas break.

The bush work day was short. After the morning milking was done, the barn had to be cleaned and the team watered before being hitched to the sleigh around 9 am. To reach the swamp we went south a ¼ mile on our plowed side road to a four-way intersection, then a half mile west on an unplowed road. Here we angled across the 10-acre back field, entered the ash/cedar swamp, then onward a few hundred yards to the balsam/spruce swamp. Dad never allowed Doll and Bess to trot, so the cutting never started much before 10 am. We took a short lunch break around 1 pm with a fire only on the coldest days. That left about two more hours of cutting time before loading the sleighs. Darkness would often catch us just before we put the horses away. There was only time for water and a rub-down while they got an oat bonus – they would get their timothy when the cows were fed after milking.

Bess and Doll having a pre-workday safety meeting.

One sleigh-load a day? We were pulp-cutting whizzers, you say? But not so fast, my friends – the loads were only slightly more than half a cord. Our team was Percheron cross, not overly large horses and our flatland sleigh was not built for rough tote roads.

We piled the pulp beside the end of the unplowed road at the four-way intersection and day by day the pile grew. Sometimes on a Saturday Walter and I would forward a load or two while Dad and Rod cut. By early April there would be 50 cords of spruce on one side of the road and 20 cords of balsam on the other with room for a truck between the piles.

Flatland Truckers on the Pulpwood Haul

The wood haul started around Easter break and I don't' think the DHO (Dept. of Hwys. Ont.) posted spring half-loading at that time. If they did it wouldn't have affected us – the 2-ton could only carry a half-load anyway. So we hauled pulp, and it was a fairly slow process.

Our wood rack was not designed for the job. The twelve-foot deck was strong, but the headboard without the sideboards to hold it in place leaned forward, touching the cab roof. The end stakes were ash pickets trimmed to fit the small pockets and thus leaned backwards. The original setup was made to haul grain and cattle, but we managed to haul our wood.

We were also laughably unsophisticated – trying to be honest in a dishonest world. Dad would never ship a dry stick and if an old spruce had a hollow butt-end we trimmed it back to where it was solid before taking the first 8-footer. This waste wood was always hauled out anyway, going home to the wood furnace.

Tops had to be no less than 2 1/2" for spruce and balsam and dad wrote that in stone, so we all carried micrometers in our hip pockets (just kidding.) Four-foot sticks were frowned upon at the mill gate but we did put some on the load – always with the big end out (ok, a little sneaky) and we packed the load like sardines. If I was on hand I was top-loader and I was always told to roll a crooked 8-footer to fill a gap – always.

So between the leaning stakes, the tight load and the scaler always taking three inches off the top of a 2 1/2 cord load, we gave the mill a lot of free wood. One small compensation was that we were seldom "docked" at the woodyard gate.

(Fourteen years later I helped Roger Soucy haul a load or two out of Wasaw Lake country. Every time I rolled a log Roger would holler at me – and he did a lot of hollering – I just couldn't stop rolling.)

Table Scrap

Cecil Kellar was delivering his own poplar with his own truck. Loads of poplar always took a hit at the gate – too many gaps, punky ends, small ends – you get the picture. One day a scaler docked Cecil eight points – eight tenths of a cord on a four-cord load and Cecil had had it!

"Mark those logs," said Cecil.

He left them on his truck, and on his way out the scaler flagged him down. "Where are you going with our wood?"

"That's my wood," said Cecil, "You won't pay me for it, and it will feed my stove at home."

Thereafter Cecil never lost more than three points.

Flatlander Logging Contractor Economics.

Pulp prices in the early '50s were excellent and unequalled to this day with inflation factored in. Spruce peaked at over $26/cord and Balsam at 23 – 24 bucks.

But we had expenses to consider. Doll and Bess ate hay and oats, which they would have eaten anyway – four men, likewise.

Dad paid no wages, no withholding, no UIC premiums, no worker's compensation tariffs, and no holiday time.

We had to licence the Ford, which was licenced yearly anyway, and I think the licence was five bucks a year, entitling us to haul half a ton! Locally there were no scales to cross and the yearly trips to the Deerhorn place (which we still owned and cropped) passed only one scale at West Hawk Lake, just over the Manitoba border. That was no problem either – farm trucks were rarely checked, and the only reason we ever did stop at the scales was to say hello to our Uncle Claire who worked there – sweet!

We never ran the two-ton into a tote yard, sticking to three miles of gravel east of the wood piles and then on pavement for four miles to the mill wood yard, so we seldom replaced a tire.

Gas was 25 cents a gallon.

Our biggest expense was swede saw blades – us flatlanders were tough on blades.

So you can do the math – we ran a conifer grow op – 99.9% profit.

Frank's Table Scrap

One day a neophyte bought a chain saw. The next day he went in to town to take it back.

"You said this thing would double my production," He told the store owner, "But now it takes twice as long to cut a cord with this thing."

"That's odd," says the store man, "Maybe it's not running right." He grabbed the pull starter and brought the little engine to life.

"Hey!" cried the guy, "What's that noise?"

However, unforeseen expenses can occur, as happened one day in early April. The spruce had been delivered and we were hauling balsam. For some reason the mill always took the balsam last. Perhaps it was because frozen balsam was difficult to peel in the drum barkers, and in early spring, even though it had been cut last winter, the sap was now oozing out and the bark would loosen more readily.

But – while this may have been advantageous for the mill it was a negative as far as I was concerned. That balsam sap gummed everything up.

You couldn't slide the logs. You had to pick up one end with your pickaroon, pull it up on the load, nail it again near its centre and plunk it in place. The heavier butt logs were a two-man operation. Then at the wood yard the unloading would be the same – a slow, sticky job.

That sap stuck to everything! The pickaroon handles soon looked like a wooden-centred balsam sap candy bars and leather mitts became sticky flytraps. It was usually warm enough to work without mitts but who wanted balsam sap hands! Walter had to steer the Ford and I didn't want to go back to school on Monday wearing balsam gloves.

One day we got a break at the scale shack. The scaler told us to take our load to the river. A short block away on Front Street near LaVerendrye Hospital there was a spot where we could toss the balsam directly into the water. The river was open now, the river chain up to the wood room was operating and the balsam would float down to join jack pine to feed the chain.

This was easy-breezy and the logs flew off the truck, I got over-enthused, buried my pick to the hilt in a big log, gave a mighty heave, and away into the drink went log, pick and leather mitts. That stick must have had ballast on the opposite side, because it popped up and floated away downstream with the pickaroon on top held by two mittens with invisible hands – funny to watch.

Some guy at the river chain got a bonus, and my paycheck was not deducted – what paycheck?

Table Scrap: Hauling Pulpwood - as Told by Stan Olson

Those were the good (?) old days on the wood haul. You bought a truck, built a strong 12 foot pulp rack and with a binder chain, "bear trap" chain tightener and a pickaroon, you were in business. You also, as an owner/operator, needed a strong back, an exemplary work ethic and a fair amount of good luck.

Like most of the truckers from the mid district, Stan hauled two loads a day – hand-bomb the wood on, drive 50 miles to the mill, and hand-bomb the wood off. He generally hauled five cords per trip at five dollars per cord – 25 times 2 equals $50 per long day.

Stan Olson and his first truck

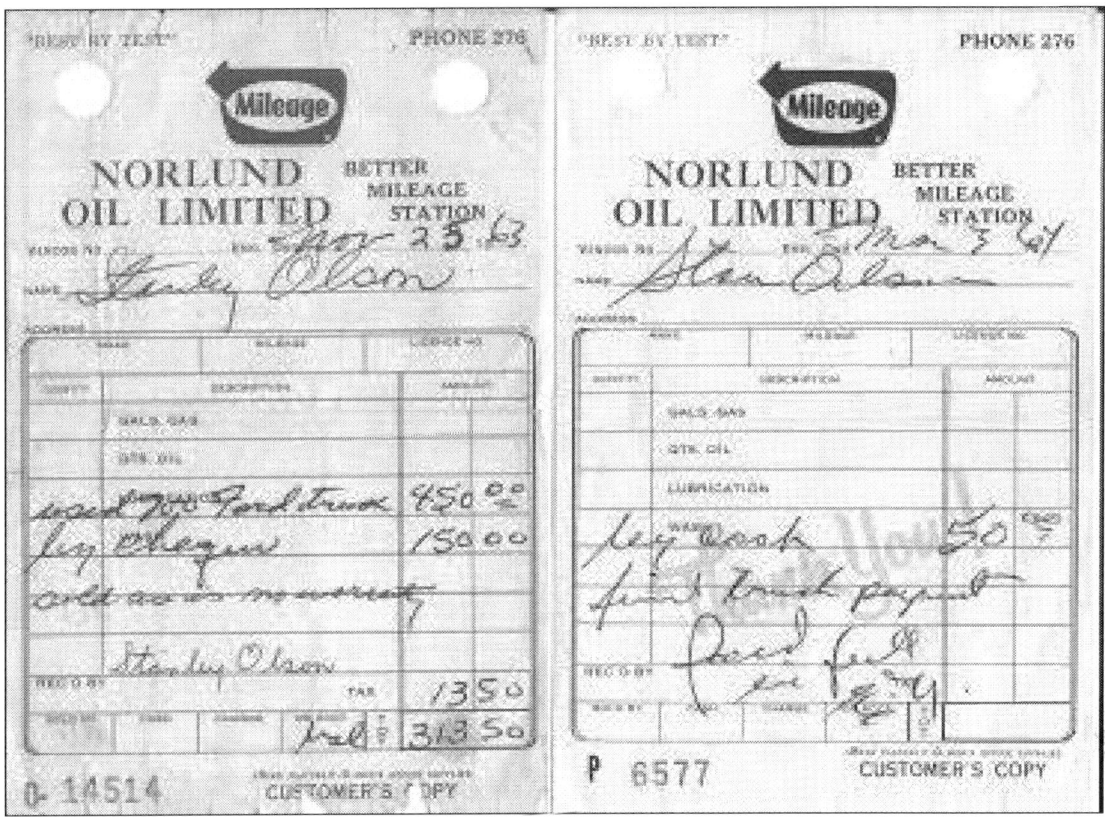

But snow was deeper back then and rural roads had high snowbanks. When an unfortunate wolf got hemmed in between the high banks and "happened" to be hit, he joined the top of the load and brought in a $25 bounty – bonus!

Many years have passed, and Stan is still hauling pulp, now pulling a five-axle trailer.

Woodyard Rules – Strange but True

The O&M (Ontario and Minnesota Pulp and Paper Co.) had some screwy woodyard protocols – very difficult to understand.

We always "hand-bombed" our loads on and off. Some trucks had homemade PTO-driven cable deals to snaffle the big sticks but they were only used for loading. No one in the early '50s had hydraulic self-loaders – they would start to appear later on in the decade. I think most of the others used our method, which was to load the large logs on the bottom with the smaller stuff on top.

Now we get to the woodyard. We were directed to an area where the wood was piled eight feet high and we would extend the pile. If we were starting from the ground up it wasn't at all hard to do, but the pile had to be eight feet high, so we often had to add our load on top. When we came to our heavier bottom logs we would sometimes cheat by moving ahead a few feet. One day the yard foreman caught us in the act and Walter got a royal chewing out, so we promised to be good. Those foremen had the power to bar us from the yard.

Here's the weird part. There was a movable dragline-type wood clam in the yard and as we worked he would be picking clamfulls of wood from farther back on the eight-foot pile building his own pile – a high one. We did not expect him to unload our two-ton, and we never saw him unload any other wood trucks, but why did our pile have to be exactly eight feet tall?

One year the O&M really confused us. We had to take our balsam to Devlin and put it into a CNR "gondola" car. This time our job was easier – we hauled it before the sap started running and because the car was on a siding we had a high, solid ramp beside the car and mostly tossed the logs downward. Of course this involved some bottom loading and we had to stand eight-footers at each end so the load would extend four feet above the gondola walls. The thing was – it didn't make sense to be hauling wood the wrong way.

Another thing – Dad was told the car had to be loaded in three days or CNR would charge O&M demurrage. With all hands on deck and pedal to the metal we made the deadline – and the car stayed put for three more weeks.

The final kick in the brisket was that Dad was not paid until the car was scaled at the wood yard and I'll double-damn guarantee that the two upright palisades were never scaled.

Flatlanders and Forest Husbandry

Dad would only select-cut. I don't know where he learned this, he had never select cut wheat and barley in Saskatchewan, but he would only select-cut, and it drove me bananas. Why, you may ask? Well – I was an early-teen know-it-all and I knew it all thanks to my friends in school. Their dads, their uncles and their older brothers strip-cut and to me it made more sense. Strip-cutting is self-explanatory. The woodlot is cut in strips and the sleigh road runs straight down each strip with one-half to one cord piles beside it. Straight down and straight back – it was simple, neat and logical.

Our sleigh road would have made a garter snake dizzy. Dad would never take anything under an eight-inch butt unless it was in the way. We also made sure that small uncut juniors were not "barked" so Doll and Bess had to go wide on the sharp corners, which also confused them. But we select-cut and I bit my tongue.

Dad died in 1974, we sold the farm in '76 and five years later I stopped in to chat with the new owner. I asked him how the woodlot was holding out and he said they had gone in with a skidder and had clear-cut more wood in one winter than we had taken out over nine years of selection. The old flatlander was right on the money.

A Flatland Incident on Manitoba #6 in 1951

It was the last two weeks of August. Walter was taking the Ford back to Deerhorn with the Model B loaded on the back, sideboards and end gate attached. (The B went back and forth, spring and fall, for three years, putting the Deerhorn crop in and taking it off until Dad sold the place.) I was 13, missed my best friend Reg Henrotte and talked (wheedled) my way into riding shotgun.

The two weeks turned out to be less than advertised. After our few acres of crop were finished, Walter had grain to haul for other farmers. The harvest moon had brought fine weather so he would be running around the clock, catching some shut-eye in the cab. I hung out at our old farm-house (now rented out) and although the folks were nice, they were all older than me. I could walk the mile west to Reg's, which I did every nice day, but I was a fifth wheel – Reg had his own farm duties to attend to. So it was pretty boring – until Walter burned the clutch on the Ford.

The issue was that although the clutch could be replaced in Lundar, the two-ton was loaded with five and a half tons of wheat. The wheat had to go to the elevator at Warren, forty miles south, and it must be delivered now. It was a custom haul and the customer did not want to take the chance of his wheat quota being cut off.

Walter came to get me and the B. My job was to be on the front end of the chain heading south. Walter figured I should handle the tractor because if I lost the truck and spilled the load, both truck and wallet would be damaged. Another issue was that if I had to snub it up I lacked pedal power – the Ford had no vacuum brake booster.

The last thing Walter told me was that if I felt I might doze off I should give him a wave and we'd take a break. I scoffed at the thought. Moi, fall asleep? Except I did with six miles to go.

I was dead sure I wouldn't drop off but the Deere topped out at eleven miles per hour and it was a nice, warm August afternoon and the next thing I knew I was in the ditch with the Ford above me on the highway and both of us googling each other with wide eyes. Walter had already hit the binders expecting me to do likewise, but not me – I reefed her back onto the road.

Three things saved my sorry ass that day - the ditch, the B and the chain. The ditch was broad, dry and smooth with no lurking boulders and no nearby driveways. The John Deere was designated "BW" meaning it had a full front axle, and though the axle had a centre pivot point, it only travelled so far until it hit a "stop." When I peeled her up onto the gravel Walter said the right rear tire was almost three feet off the ground, but when I hit the end of the chain it brought the tractor back on all fours before a link snapped.

We tied a knot in the chain and I only blinked once for the next six miles but I still don't understand why I was replaced for the return trip to Lundar.

Flatlanders Moving Machinery

We had moved over Christmas 1950. The Model B. our hay wagon and Massey mower came down in May or June of '51.

It was decided to kill three or four birds with one long stone. The rubber-tired farm wagon and hay rack were new and it and the Massey were needed in Crozier. So with the wagon hooked behind the mower which was hooked to the B, Walter set forth with numerous harrow sections, and other such stuff piled on the rack.

To put it mildly it was a looong trip. He went south on #6, cut east on country roads to cross the Red River at Lockport and picked up the west end of Manitoba #4 (now 44) and on past Whitemouth, Rennie and West Hawk lake to where #4 joined Ontario #17. After going through downtown Kenora he hit the top end of #71 and continued south to Kereliuk's Corner where pavement ended at that time. He now turned east on gravel to Halvorsen's Corner and south to Emo, where he hit pavement again on Hwy 11 to our Crozier side road.

At "B" speed the trip took two long days with an overnight stop at Rennie, As a boredom reliever when he hit hill country he let the train freewheel downhill, a relatively dangerous thing to do. Even though the B still had tight steering response, he couldn't use the individual brakes at high speed, so he just let 'er rip.

On 71 he picked up a couple of army lads and they rode the rack. On a long hill he did the Georgia overdrive deal and the train started to fishtail with the wagon at the back cracking the whip. At the bottom the boys gave him the high sign and they got off, saying they would rather walk.

The stretch south of Halvorsen's Corner had little or no shoulders. It was dark by then and because the B had no low beams to give oncoming cars a break, Walter would shut the lights off before they met. He got too close to the ditch and dropped a front wheel which he pulled back right smartly. He arrived home at 2am and in the morning an inspection showed one right front axle covered in reeds and swamp grass – close call. And he never fell asleep at the wheel – I could have done that for him!

Flatlanders Tackle the Muskie Bay Hill – Late March 1955

I went to Winnipeg in the two-ton with Walter – the purpose of the trip being threefold. A load of cull cows was going to the stockyards and the backhaul was another load of newly purchased bred heifers for our milk herd. I would be bringing back a pickup – a '53 Ford short box.

I got my orders before we headed home. I was to follow Walter, the excuse being that if the '45 broke down I would be on hand. I could read between the lines where it said, "Keep an eye on that sixteen-year-old behind you."

I don't recall whether Manitoba # 1 was finished yet, but it really wouldn't matter if it was – 30mph was our max. What with loading the cows and all we got a late start and it was coming dark when we went through Kenora, and it started to snow really hard. This was a typical spring deal with huge wet flakes coming straight down.

Walter pulled over at the 71 junction. I walked up to see if he had a problem, which he did. The '45 was well past its best-before date and the vacuum wipers couldn't handle the soggy snow. He was driving with his window open, using his left hand as a windshield wiper and he had eight inches of un-snow-covered glass to navigate with. This would slow him down even more and he knew I would have trouble on the hills. He was loaded, I was empty and had four bias ply summer tires, so he told me to go ahead, be careful and wait for him at Finland, where the hills ended.

I drove on just fast enough to make the hills and slow enough to make the corners. Forward vision was no problem – I had electric wipers and soon Walter's headlights disappeared from my rear view mirror. There was no other traffic – zilch – and I did ok until I hit the Muskie Bay Hill.

Even now that hill is long and steep, but in 1955 it was longer, steeper and higher. The highway still turns left at the bottom, climbs and flattens out a bit before turning south for the second ascent, but in '55 the bottom turn was tighter and steeper. This wasn't my first Muskie Bay rodeo so I knew what was coming.

By now there were three inches of heavy, wet, easily-packed, slippery snow on the road. A decent straight stretch led to the first left-hander, but I knew I couldn't run the grade. If I hit the upward curve too fast I would slide into the guard-posts on the right. This deal was double-difficult. The highway was banked on the curve, and if my ass end swung sideways I'd be a sitting duck for northbound traffic. I hadn't met any oncoming yet, but you never know – maybe some other goofball might be on the road tonight.

I hit that curve as fast as I dared, made it less than half-way and spun out. I backed down to the straight stretch, gave her another shot and made it another ten feet. I had two hundred yards to go to the midway mezzanine and it looked like the snow would melt before I got there. This time I backed up the full half-mile before starting my run.

I only gained a few feet this time. It was a bit steeper here, so I decided patience might be the answer. By holding her slightly above idle the summer treads found their way to pavement and I moved a head two feet at a time, but I slid left somewhat on each bite. I had to stay with it, and ever so slowly I side-winded along until I was on the wrong side with my left wheels on the wrong shoulder near the wrong guide-post wires.

This worked out better than I had expected. With the weight transfer being to the left, and with gravel under the snow on my left rear, I not only made slightly more progress, but I also stopped drifting left.

Uh-oh! Headlights were coming – an early warning reflection on falling snow. I flashed my lights and held my breath. When they appeared around the corner above me I could tell they were moving at a snail's pace. Headlights followed headlights, and the first unit went slowly by. It was a Jeep – an honest-to-goodness four-wheel-drive Jeep! It was army brown with a canvas top, army insignia on the door, and with a tiny flag on a tiny flagpole on one fender. Behind it followed canvas-topped one-tons, likewise dressed-up six-bys, a couple more Jeeps pulling field-piece trailers and with another Jeep rounding out the convoy. I was elated – there were enough non-directional treaded four-wheel-drive tires to pull me straight up the side of a rock cut – and the peckerheads passed me by with barely a sidelong glance.

I was pissed and deflated, but what could I do? Then shortly after they disappeared Walter went by – the pokey two-ton had caught up to me. I hope he wouldn't stop, and he didn't. With a load on he could make the hill, which he did, stopping at the crest. I was almost over the first hump, anyway, gaining more ground with each burn-down. At the mezzanine I returned to the right-hand side and picked up a gear, which I lost as soon as I hit the right-hander – the steepest part of the hill, but I knew Walter would walk back to help me.

Another set of headlights were coming, and he stopped. It was a '50 Chrysler and I knew the car and the guy who owned it. Herman Anderson and Jack Busch got out and started pushing, soon joined by Walter, who had walked back almost a half-mile. With three men on the back we made steady progress and when the hill started to crown the Chrysler boys waved goodbye, Walter added his 190 pounds to the back bumper and shortly after I parked behind the two-ton.

I took a deep breath, stepped out the door and my legs felt like rubber. I was exhausted, whipped and I felt like I'd pushed the '53 up the hill myself and my finger muscles were complaining – I'd had a death-grip on the wheel. It didn't last long – youngsters can replace used-up adrenaline quickly.

We had a short conflab before hitting the road again. We had only one more steep hill ahead where 71 climbed alongside Crow Lake, but it was a straight shot. Our concern was how deep would the snow eventually be? We might have to shut down the little Ford and come back for it later.

There was no point in worrying, so we jumped in and got right at 'er. Right off the bat the white blanket on the pavement seemed thinner. Halfway down the other side the pavement was bare and overhead the moon shone and stars twinkled – clear as a bell! It was the wee hours before we got the cattle off-loaded and hit the sack, but I didn't care – I knew I'd get to sleep in tomorrow.

All-in-all that was a pretty good trip – actually quite valuable. For one thing, we now had a decent pickup to carry milk cans and other stuff. (The Hudson was a pleasant memory now.) We had also brought home some needed improvements for the dairy herd.

(Dad continually improved the herd and over the years he increased milk production while decreasing the milk cow count. In 1951 we had milked up to 22 head and often had to buy some milk from other farmers to fill our four-can contract. By the time Dad retired the dairy business in 60/61 he was easily filling a six-can contract milking 12 – 14 cows.)

Sidebar: I learned a thing or two on that trip, one being about meteorology/topography. That slightly higher east-west trend including the Muskie Bay hill held back the low clouds, thus the dramatic change in the weather. I also learned from Dad that when on manoeuvers no one is allowed to pull out of the convoy without a darned good reason. It was up to the head peckerhead to call a halt and I'll cut him some slack. Walter had met them near the bottom of the hill and they probably figured he would rescue me.

The Emo Fall Fair

I attended my first Emo Fall Fair in 1952, having missed 1951 due to unforeseen circumstances – meaning I didn't foresee far enough ahead to prepare a good wheedle. When school started in September '51 I had to listen to others brag on it, so I made darned sure my wheedle was tuned up in '52.

I liked it right off the bat – my first encounter of the fair kind. Because I was a kid, I got the early afternoon experience and I loved it – the smells, the sounds and the sights – I loved them all.

I ran into my neighbour Tommy and we both had slim bankrolls, so we picked our spots. We checked out the ring toss where prizes sat on square boxes, the smallest and cheapest on the outside of a circle. The inner concentric circles went up tier by tier, and near the centre, the prizes were replaced by numbers, with those prizes being displayed on a high rotating shelf which the mouthy fast-talker who ran the deal would turn this way and that to show some young fella what he could win for his hunny bunny. The big prize box sat in dead centre and the barker strode around between sucker tosses showing how easily the ring fit over that box. He carried the rings on his left forearm and we noticed that his display ring always went back on that arm but was never offered to players, so we passed that one up.

We checked the games out – all of them. The dart and balloon deal looked interesting until we figured out the darts were too dull to bust a soap bubble unless it hit dead centre. And once again, we two sophisticates saw that the demonstration dart was never given to a potential balloon-buster. We kept our cash in our jeans.

How about the pyramid of white-painted lead milk bottles? (We could tell they were lead because most of the paint was missing.) But Tommy scoffed at that. He claimed that the bottom bottles were glued to the base, and I sort of believed him. Privately, I thought that he had no more confidence in his throwing arm than I did – neither of us would ever be mistaken for Willie Mays.

Ok, here's the sledge-hammer-ringing-the-bell contraption. How hard could that be for fence-post pounding farm boys – but once again Tommy tossed cold water on the project. The guy had a pedal under the grass at his foot – he would only allow the ball-bearing in the tube to hit the bell if the striker had a chick to impress.

We were running out of options. I liked the look of the Crown & Anchor wheel, but we were chased away right smartly – you needed more whiskers and more than dimes to play this one, so we settled for the rubber duck circulating brook. We laid down a dime each, picked a rubber ducky and we both walked away, each carrying a brightly colored pinwheel. They probably set Conklin back five cents a dozen, but we had outsmarted them, by golly.

Now we hit the cotton candy, next door to the candy apple booth. We liked them both, but it was a toss-up which to buy first. Cotton candy lasted a long time, but it destroyed your taste buds. Carmel-covered apples tasted great, but if you had a cotton candy chaser you'd end up with cotton candy whiskers around your mouth. It was a case of either/or – we settled for cotton.

We checked out the side shows with their colorful, possibly slightly exaggerated paint-on-plywood false fronts. This stuff was interesting, but certainly didn't fit our budget, so we never did get to see the Fattest Bearded Lady in the World, or Gyro, the Head-Spinning Man!

Last on our Conklin show list were the rides and once again we looked but did not touch. The seven-story high Ferris Wheel and the Double-Loop Love Seats looked puke-worthy, but were out of reach financially, and the other rides were "kid stuff."

You might think it was a downer, what with Tommy and I having shallow pockets, but it wasn't. We had decided over an hour ago to save our last quarters for hot dogs and pop, and cruising the carnival atmosphere was fun. We neither noticed nor did we care that the awnings were a bit faded with patches here and there, or that the Fattest Bearded

Lady was missing a plywood elbow. Conklin and other travelling midways always started the summer freshly painted, but by the time the Emo Fall Fair visit rolled around, they had been torn down and moved many times. In two weeks they would do their Labour Day Weekend Grande Finale somewhere and go back to the hanger to be rejuvenated for next year. So not to worry – we still had an hour to kill and the best stop of the afternoon awaited,

I had been catching glimpses all afternoon, and I could hardly wait. It sat in a large grassed area east of the Ferris Wheel – every farm boy's dream of unattainable riches – the tractor and machinery display. There was no hurdy-gurdy music here, no fast-talking front men, and no faded or chipped paint. Everything looked new and shiny and smelled new and shiny. John Deeres, Allis Chalmers, Massey Fergusons, Fords and others, with some tractors hitched to new mowers, plows and side delivery rakes. All of them pawing at the ground, raring to go home with me and make farming easier.

I tried out the seats on every one – both hands on the wheel and my shoulders right back – don't mess with this lad or I'll plow your sacroiliac. One tractor even had a fender bracket which held an overhead umbrella – top-of-the-line operator comfort. But the star of the whole show was a '37 or 38 Chevrolet two-ton truck.

It must have once been someone's pride and joy, and it looked brand-new. It had recently been painted a medium green with black fenders and it had new seat upholstery, a spotless instrument panel and unclouded window glass. The 7.50 X 20 tires were also new and the rest of the deal was so new and innovative that it blew me away.

A farmer (I can't recall the name) had turned it into a hay hauler – and not any old hayrack was this, it was a field-to-hay-yard one-man operation! A frame extended beyond both rear cab corners and arms came forward beside the cab. Attached to these arms was a "sweep" which doubled as an overshot stacker. The operator drove around the field picking up hay bunches and the sweep lifted them over the cab to deposit the loose hay on the deck. How the deal was motivated I cannot recall. There were no hydraulic cylinders and I think an extra PTO-driven truck axle drove a cable winch which somehow did the job – but there is more to come.

Once the load was on, the guy had to unload it – right? Well, that was also taken care of. A series of parallel bale elevator chains ran in grooves on the deck powered by sprockets which were also PTO driven. At the back of the deck was an eight-foot ramp which in travel mode acted as an end board. At the hay yard the operator unhooked two side chains, dropped the tailgate, and the deck chains unloaded the stack. What a dream! I wish I now had a real photo rather than a mind picture.

One important issue I cannot recall. Did the deck also have tubular steel sideboards? It must have had these, and if so, the stack, once on the ground, would have had to be top-dressed to shed rain.

The next time I saw that truck was two years later. It was winter, the pulp haul was underway and I met her riding home from high school with Rod. Along came the old Chev, and she was hauling pulp! It was so sad to see her and she also looked sad. Her grille had taken a blow and a fender carried a dent. She soon disappeared from the haul and I was happy with that – she had looked so dignified, all dressed up in '52 – now she looked like she'd been rode hard and put away wet.

The thing was, she had already been outdated in 52. Everyone was going the square bale route and we would soon be putting our own loose hay slings in storage ourselves. I hope she had a decent send-off, or perhaps she's waiting in a barn somewhere – what a find that would be.

The Princess and the Clown – 1954

Before I turned 15 I became a member of the Crozier/Devlin 4H Club chapter. I was not eager to join but Mr. and Mrs. Captain talked me into it – in fact, insisted that I sign up. They thought 4-H would build "character," so I assumed I was lacking in that department.

It turned out to be interesting fun. Mr. Stewart from Roddick was our leader, and he was a nice man. Few of my club-mates were familiar – they were from South Crozier, Roddick, LaVallee and Devlin, but they were all OK guys. Meetings were rotated and I only attended those that were held in our North Crozier Hall. At these meetings I learned more about beef cattle than Holsteins – the others all had steers to show at the Emo Fall Fair. I was the only dairy kid in our group and would be competing with kids from Emo and westward. But cows are cows, yearlings are yearlings and show protocol is show protocol, so I listened and learned.

We were taught how to present the animal and ourselves. We all made fancy halters using new rope and we spliced, braided, and with a false start now and then, each 4-H kid made two halters. One was for rough halter breaking and the other for show – I was only able to attend enough meetings to make my show halter.

I don't know when the others started halter training. I started after the last of the winter snow. After school was out I heard the others had some group practice showings. I was always invited but no one offered to take me, but I didn't care. I didn't need a practise halter, I didn't need practice showings, and I didn't need to learn how to wash and trim rough edges of animal hair – I had Princess.

She would be nearing twelve months by Emo fair time and she was purebred from the tip of her black-and-white nose to the tuft of her black-and-white tail and from the top of her straight back to the bottom of her perfect hooves. Her brown eyes showed curiosity, empathy, kindness and intelligence. She would be registered this fall and although I can't remember the name that would be on the document it was a long one, as befits a princess. Before September rolled around she would prove that she had far more sympathy, smarts and "character" than her human handler.

Halter breaking was easy once we humans figured it out. Princess was put into a headgate, introduced to the halter and she didn't like it, twisting her head this way and that – quite agitated. We decided to leave the halter on, leave her in the gate for fifteen minutes and try again tomorrow, which we did with the same result.

On day three we decided to try her out "free range." I got a death grip on the halter rope, Rod opened the headgate and stood by in case I needed help. Princess backed out of the gate gave her head a shake, looked at us with a "Now what?' attitude and fifteen minutes later she had the follow-the-leader concept down pat. It was the headgate she didn't like.

She never did fight the halter. She lived in the loose housing barn addition, and when I opened the door she came right over – she liked her 20 minutes of freedom. When the herd was put out to pasture two weeks later she joined them and gladly came in with the cows at milking time. She knew she would get a treat after each day's walk-about.

She soon became my buddy – my constant companion. When I went to bring in the herd for evening milking she trotted to meet me, got an ear scratch and walked back to the barn at my right side – as if she carried an invisible halter. The walk-about exercises always lasted no more then 20 or 25 minutes, and she called the time out. Her inner Holstein clock told her (and me) when to shut down.

Emo fair time was coming and the final dressage was easily done. No electric clippers were needed – we had no power anyway. She got a scissor job around her horn nubs, ears and tail and a brush-down. There was no rain the week before the fair so she had no mud to walk through and she and I were ready to rumble.

We would show our animals on Saturday. Mr. Stewart had a homebuilt cattle trailer and he would bring the Roddick steers to Emo, stopping at our place to pick up Princess. Brother Rod took me to the fairgrounds early and I hung around the 4-H area, my main concern being would I show well – I knew Princess would.

Mr. Stewart pulled in, and as soon as I saw him he saw me, and I knew I was in trouble – they had forgotten to pick up my heifer. I'm not sure who felt worse – me or Mr. Stewart. He was so sorry and I was sorry, but sorry dogs don't hunt. He had no time to go back for Princess. He had to get the other boys ready, so maybe I could help out.

I hung around for a bit but once again I was a born-loser fifth wheel, so I wandered over to the farm equipment display, sat on a new tractor, and considered my options, which were ZERO.

I couldn't phone home and they couldn't call me. There was no phone in the 4-H area and even if I found one I wouldn't know the drill. We were still on the party line hand-crank system at home, it was long distance to the Fort Frances exchange and I had no clue about calling collect. By this time Dad would have figured out something had gone wrong, but what could he do? The two-ton side boards would have to be installed, a make-shift loading ramp scabbed together and I knew that there was no way, no how, that Princess could arrive in time. I wandered around watching the midway wake up and when the 4-H show ring started I checked it out. I didn't stay there very long – it was too painful to watch, so I "bought" farm machinery hoping Mom and Dad would come early to take me home, which they didn't, showing up at three or four pm. By that time I couldn't go home – I had one last duty to perform and someone would have to fetch me later.

I couldn't believe it – I had to chip in at the 4-H food booth between five and six pm. All the clubs had to do their hour and Mr. Stewart had run me down earlier to shanghai me. I didn't have the cojones to refuse but I had a couple of hours to think about the injustice of it all and to plan my revenge which would be a sharp stick in somebody's eye. It was a good plan, a well-thought-out plan and revenge would be sweet (at first.)

The five-six stretch was a busy time. We sold mainly hotdogs, fries and pop. For every pop I handed over to cash customers, I drank two, and hot dogs were also a two-for-one deal. The other kids didn't say a word, but one of them must have told Mr. Stewart who soon showed up. I looked him right in the eye while stuffing an entire hot dog into my mouth, and he turned silently away.

I was a man on a mission until the mission was aborted. I went behind the booth and horfed up the whole works – puked until my stomach was sore. I sat quietly in front of the ladies' vegetable, crafts and such pavilion until Walter came to pick me up at seven. He suggested some midway time but I just shook my head and I didn't talk much on the way home.

The next day at home was also pretty quiet. I was not about to point fingers at Mom, Dad or Mr. Stewart or myself. We were all guilty, so I figured I would be a Spartan Stoic, able to handle the blows of life. I had everything under control when I went out to pick up the cows for the evening milking.

As usual, Princess saw me coming. She trotted over and gave me a little head-butt, expecting a noogy before she walked home at my side. That little loving nudge was all it took to knock me off the "I don't give a rats's ass" mountain-top. Of course there was someone to blame – it was all her fault! And I went Berzerkerstan! I found a stick and nailed her a good one on top of her head, followed by a few whacks on her flanks. I chased her, calling her a stupid, stupid cow – screaming at her until I was out of breath. Then she turned and looked at me with those soft, brown eyes – no judgement – no recrimination – just a "What did I do?" look. If she had been angry I could have handled that but this was way too much!

I found a rock to sit on and I bawled like a baby. Along with the tears, years of frustration poured out; years of unkept promises, years of listening to endless screaming arguments about money, and also remembering how, when I was nine, my boots were too small so I curled my toes rather than ask for a new pair – it all came out. Just when I thought the well had run dry, Princess came back, kissed my ear, and the stream found new water to flow. The last spate was soothing – I hugged her neck, asked for her forgiveness, dried my eyes on her soft coat and we went home friends again – what a Princess!

That was a life-turning point for me. Although it was not a conscious decision on my part, I started to lose interest in the farm. Over the next five years I would move away figuratively, sometimes literally, and it was sort of a Chinese torture routine for me and my family until Rod backed me for that last stab at high school in Trenton.

Princess and I slowly separated also. It was not at all a bitter parting and she still walked with me until winter came, then she went back into loose housing, eventually became a mother and the top producer in the herd – much more successful than I.

One sunny afternoon in October Princess wore the halter one last time. They were filling out the registration papers at the kitchen table and I was totally disinterested, but Mom said I should draw princess's markings on the forms. (One page had a blank outline of a cow – broadside left and right.) I could do them from memory, but they insisted I lead her out to pose. Accuracy was important.

Princess was happy to oblige – proud, even! Maybe she hoped we'd get back to our daily walk-about, and she posed left and right, standing perfectly still while my pencil snapped the last photos of our beautiful friendship, and the halter was hung on a hook, never to be used again.

The Year We Cut Fort Frances Down – Summer 1953

It seemed we were always short of hay back then. We were a bit shy on hay land, especially if a field was in crop rotation, but part of the problem was our old-fashioned haying method. In '51 and '52 we put up loose hay. When the hay was cured and bunched, we brought it in by team and wagon to the barn. We had loaded the hay on slings – three layers, like building a Dagwood sandwich on the rack. Then they were hoisted to the loft by a complicated rope/carrier dealie and the hay was spread to distribute the weight evenly. The loft hay was fed to the horses and milk cows – stacked hay was for the young stock.

Tame hay should be cut at its optimum protein content. Our system was pokey, so some hay was cut too late. Added to the mix was rain showers – after all, this was the Rainy River Valley, so half the time protein would be lost while waiting for the hay to dry. Yet we were very careful about this – damp hay in a loft can heat and spontaneously combust. (Spontaneous combustion – two ten-dollar words I learned very well in the early '50s.)

So depending on quality and quantity we generally ran out of hay before spring. Hay had to be bought, cutting into milk profits, so Dad made a deal.

We shipped our milk to Flinders Dairy. Back in the day. O.B. Flinders (Grandpa) had a farm, milked cows and delivered the milk to town folks, which naturally led to a now fully-modernized dairy. He still owned most of the land across the highway from the dairy with the exception of highway lots he had sold. McCool's Mileage Station was near the Frog Creek Road, Smitty's Drive-In was being built further east and there were a couple of private houses. Another lot was being prepared for the first Dairy Queen just before West End Motors. The rest had sat unattended for a few years now, and maybe the town was bugging O.B. or maybe he just wanted to spruce thing up, but for whatever reason, he asked Dad to cut the hay – no charge.

This would be a daunting prospect for us. There was almost two hundred acres to be cut and hauling rackfulls of loose hay six or seven miles was out of the question, so we modernized. A side delivery hay rake was purchased, and Mr. Bragg would custom bale for us with his Minneapolis-Moline tractor and a MM wire-tie baler – a matched set.

We baled the tame hay at home first, and it went lickety-split. The town two hundred also came off without a hitch. There were a few years of old bottom and some small willows mixed in with new growth, so good weather cured the hay rapidly. I did my share of tractor-time, sitting up there ramrod straight so those town boys could tell how important this cool dude was. Fat chance – the only wave I got was from a train crew, and they'll wave to anybody.

We hauled the bales home with the two-ton, leaving the field neat as a pin from Frog Creek Road to West End Motors and from McCool's back yard to the CNR Tracks. Yup – we cleaned Fort Frances' clock that year.

And man - were those wire-tie bales heavy! Our hayloft was full of tame hay for the milk cows, so the FF bales were piled outside and were fed to the heifers, steers and non-milkers. They would pick out the good hay and leave the rest for bedding, and this was another bonus – we had no straw that year. Triple bonus – by spring we had enough haywire hanging in the loft to repair machinery for many, many years.

There was one bummer in this baled hay deal – it fouled up our annual camping-out trip. For the last two years when the last of the loose hay was put in the loft, Judy, Frank and I would haul blankest and quilts up there for a sleepover. We'd snuggle into the sweet-smelling alfalfa, clover and timothy, cover our heads to escape pesky mosquitos and tell ghost stories before dropping off. Mother nature always co-operated. Later on we would be treated to a light and thunder show followed by raindrops on the tin roof – wonderful!

We were stubborn enough to try it one more year, but square bales are not exactly Sealy Posture-Pedic mattresses, so we packed up our quilts and went back to the house.

Sidebar: That was the only year we had Mr. Bragg do our baling. Wire-tie was expensive and Dad bought a Cockshutt twine-tie from Nick Rogoza in Devlin. Decades later I learned that Mr. Bragg dealt with logging camps. Those heavy wire ties were loaded in boxcars and went east to Mine Centre, Crilly et al, where loggers like Beaver McEvoy and the Stewarts had horses. The wire bales never broke open when manhandled, and they were handled many times before they reached the bush horse barns.

Hockey Night in the Rainy River Valley – Eastern Conference.

Hockey was hockey back then, and hockey night was Friday or Sunday afternoon – no one even thought of competing with Foster Hewitt. R.R.Valley hockey was played on outdoor natural ice rinks, and the Fort Frances Canadians played on natural ice in the Old Barn until the Memorial Arena was completed just in time for their final Allan Cup series in 1952.

Singing the National Anthem in the new Memorial Arena.

During the Mini Ice Age of the '50s the natural ice went in early and melted late. Mr. Hampton had a GMC 3-ton and either owned or borrowed a 1000 gal. water tank, so he was the "Ice-master." Starting at freeze-up, three or four loads would lay the base and another load or two of steaming hot water smoothed out the ripples. The hot load was always donated by Flinders Dairy and P.F. Hampton donated the GMC time.

It was a four-team circuit – McIrvine, Crozier, LaVallee/Devlin and Emo. They were all pretty evenly matched with a mixture of solid veterans and hot-shot rookies. I can only recall that Dean Whalen played for McIrvine, but I'll always remember their rink. One end was at the edge of a creek and had a high, chicken-wire puck catcher. One night a deflection sent the puck high over the wire to be swallowed by the ravine and the game was over. No one had thought to bring a spare hockey puck.

LaVallee/Devlin had a pretty good team, always staffed by a couple of McTavish boys, and a Cain or two, and they were our main competitors. Emo? I don't remember one single player from Emo. Emo was waaay out west – you had to pack a lunch to travel 15 miles in those days. However, I do remember that they had the best rink. We could sit behind plate glass at one end in heated comfort.

Of Course, Crozier was "da Bes." We had 'em all. We had high schoolers – Allan Gustafson, Dennis Robinson and a couple of DeGagnes along with some older young fellers like George Hampton, Tony Wier and Norbert Bragg. Thrown in were some wily veterans – Walde Kliner, Mike Wodell and the Stewart brothers, Nels and Ray. The Stewarts were fast, smooth skaters and sometimes played with McIrvine. Quite often players would change sweaters (What sweaters?) when work or circumstances left a team short-handed. The Stewarts were almost McIrvinites anyway – McIrvine was more country village than city back then.

We threw in the odd ringer once in a while. One winter our future brother-in-law, Leon Brandson was imported from Lundar, Manitoba. He was apprenticing at Stewart Motors and was a hard-nosed defenceman. Take a mild-mannered Icelander, + a pair of skates, + a hockey stick = instant animal!

A couple of Walde's brothers gave us a game or two. Bob Kliner showed up one night with his goalie gear. He played on an army team, was home on leave, and nobody scored on us that night.

"Torpedo" Ed Kliner also put in a couple of shifts – cameo appearances. He was playing for the Fort Frances Canadians so he was an unmarked man and he was so fast no one could hit him anyway. He was also a semi-legend locally, having had some NHL experience, so when he broke his stick and tossed it over the boards, my big brother retrieved it for me. It was unsigned, But Torpedo Ed had used it, broken it, and I kept a broken hickey stick for two years before I gave my head a shake.

About the time our ice started to melt things were heating up in Fort Frances. The Canadians were on a roll towards the Allan Cup. This is well documented locally and on line, but my take as a thirteen-year-old may be worth examining.
One thing you won't find or may not know was that their first series vs Fort William was the toughest. The year before in 1951, FW had taken a close series from FF en route to the finals before losing to the Owen Sound Mercurys.

This year the Canadians had lost a premium player, Morris Saplyway, who decided his own home town, Fort William would, with his help, take all the marbles. It looked like it would happen, because with speedy Morris leading the way, Fort William led the series 3-0 until Fort Frances gumption kicked in and won the next four games.

After pesky FW was dealt with, FF handled the western Canadian teams quite easily and the final series was against the Stratford Indians. – East vs West.

It may or may not have been the final game, but Rod and/or Walter took me and it was grrrreat! I think the fire marshal left town on purpose because the new arena was packed! Some folks with seats had not-so-young kids on their knees and standing room was squeezing room only. I was not yet full grown, so I was allowed to hang on to the rail and hang on I did. Everyone had to take turns breathing – a collective inhale would have surely bulged the walls.

What a show! We had all been hearing the names on CKFI radio broadcasts, but now I could put a face with the name: Ike Eisensoph, Mike Hupchuck, Verny O'Donnell and many more – if pressed, I can still name most of them. Dun Sampson on defence was an International Falls lad, but he was married to a Gustafson, giving him diplomatic immunity. Sambo Fedoruk had a booming slap shot – and had we ever heard that term in 1952? No matter – it sure boomed and Dennis Robinson said, "And he wonders why his stick looks like a broom."

The star of the show was Bruce Dale – the Stratford goalie. There was a long interruption midway through the third period. The new scoreboard/time clock had an issue and Mr. Dale gave us a demonstration of figure skating in goalie pads – the crowd loved it!

What was remarkable about that Allan cup win was that all our players were home grown except for Bill Cleavely, our goaltender, who, although from 'back east", had been posted to FF with the Lands & Forests, making him a legitimate resident. Some teams, the Trail Smoke Eaters, for instance, brought in semi-pros, supported by Cominco, who hoped to add Allan Cup silver to their lead/zinc inventory.

I kept the Cup edition of the Fort Frances Times for 62 years, and even had it laminated. I gave it away two years ago to Gary Judson, who collects stuff and cares. My family thinks the cup is something Uncle Alan keeps in his jock strap.

Table Scrap: As Told to Me by Al McTavish.

The McTavish farmhouse sat well off the road, so they had a long driveway. It had snowed and blowed all day Friday and on Saturday morning their driveway was covered with fourteen inches of snow not counting drifts. The snow plow cleared the side road and did not turn in towards the house. It was Saturday morning and the Fort Frances Canadians were playing a Saturday night home game.

Everybody grabbed a shovel and eight (yes, he said eight) hours later they were off to the Good Old Hockey Game.

Franks Take:

I was too young to remember anything much about those golden years, but in 1970 I joined up to play Liniment League hockey. One of our opposing teams consisted of mostly old geezers in their '60s – and I soon learned respect for those guys – they would let you win once in a while to keep things going, but you couldn't beat them. I even went to watch them play other teams just to try and figure out a defense. (I played left defense.)

It went like this: for two or three games they would lose by one goal, then they would get your team to wager a case of beer or two on the outcome of the next one. That next game, Vern O'Donnell, small, wrinkled and hunchbacked, would grab the puck off the faceoff, pop it back to the right-winger and head for the right-hand corner at the other end. The right-winger, (any old recruit would do) would advance if he could, then pass it over to Whitey Christansen, coming up the left-hand side. By this time there maybe was or maybe was not anyone between Whitey and our goalie. If there was, he'd thread a shot past the defenceman and into a corner of the net, or else put a soft pass out front where little Vern would slap the puck past an out-of-position net minder.

The goalie didn't have a chance. If he defended one position, the puck would come from the other. Once and once only, I defeated their little dance by pretending to rush Vernie and instead stepping between him and the soft pass. The next time I tried that, I found myself all alone heading toward centre ice, circling back just in time to see Vern lifting the puck gently over a sprawling and very angry (at me) goalie.

It was best just to buy them a couple of cases of beer now and then.

Powering Up in Crozier

Electricity was coming, and electricity was in the air! Every day for two or three days, poles were being planted along our side road, with the last one at the end of our driveway. Shortly afterward, shiny Hydro lines travelled cross-arm to cross-arm ending at the transformer on the last pole, and we were ready!

Two straight cedar trees had been cut, erected between the house and barn and wires had been strung. Yard lights sat high on each pole – large bulbs shielded by a concave enamelled dish – white on the underside and green on top.

At the house we only had to connect to the outside bar. The fuse box in the basement had been waiting for four long years. The barn wiring was very simple. A single wire led down the centre aisle ceiling with five porcelain fixtures. Another wire led to the vacuum pump, (where a small electric motor would soon replace the stationary gas engine,) and continued up to the hay loft to power a 100-watt bulb in an explosion-proof globe. Everything was up to date, modern, and believe me, the early October main-power actuation ceremony was attended by thousands of cheering Durnins.

The changes this made in our lifestyle were so far-reaching and so numerous that I may not remember them all, but I'll try.

That first nice evening after milking was over and

while adults relaxed in the house with Dad reading, Mom knitting and Rod doing his homework at the kitchen table, all bathed in actual glowing-filament light – Judy, Frank and I hit the bush! The path to the barn led through poplars with bushes here and there, and we had been waiting for this opportunity. We flicked the yard switch on and played hide and seek under those amazing lights. Mostly we just dashed around, hiding in bushy shade, to pounce on unsuspecting Little Red Riding Hoods and Big Bad Wolves. We would have stayed out there past midnight, but old folks need their sleep. Two warning flicks of the switch and we were off to bed.

However, those yard lights cut into my evening cardio workouts. Pre-electricity, after milking was done I went to the house alone while Dad stayed to finish up. He wouldn't let me carry a lantern – I had two milk pails to take to the house to be washed, so I was the family hundred-yard-dash record holder. I had to be fast, because wolves, vampires and mountain lions lurked. Now, with oodles of bright lights to keep the goonies at bay, I could saunter – still whistling, mind you, and with an eye peeled to the rear.

The battery-operated Marconi went into the attic, replaced by a plastic Zenith from Eaton's.

Dad still controlled the evening airways, but after he went to bed on Saturday nights, Rod could take the radio upstairs. We would snuggle into our beds and tune to country music – the WWVA Jamboree in Wheeling, West Virginia, the National Barn Dance from Shreveport, Louisiana and of course, from the Ryman Auditorium in Downtown Nashville, here comes the Grand Ole Opry! Ernest Tubb, Skeeter Davis,

Wilma Lee and Stoney Cooper – I had no favourites, although a good yodeler was interesting. Grandpa Jones and Cousin Minnie Pearl cut no ice with me – I liked Lonzo and Oscar better.

It's hard to tell which station we preferred. We listened to them all and if the signal weakened we'd switch back and forth. Eventually we'd drift off with the radio fading to faint static before it was shut off.

So many things changed, all for the better. We were not unique – I'm sure every farm family that had powered up in the '30s, '40s and '50s shared these experiences, but young folks nowadays could never understand how much difference it made in our lives.

How about Mom? The wood range was replaced by a four-round electric stove with a reliable oven! No more hassling Dad for decent stove wood and no more wood box for us kids to fill. There was one drawback, however. We no longer had a warming oven to dry our mittens in snowball weather.

Mom could now iron clothes with a nifty new Proctor-Silex steam Iron. (Drawback #2 – I had to learn to iron my own shirts and slacks.)

Every day something new appeared. We came home from school one day to find a new refrigerator and chest freezer in our large kitchen. Both of them shining white and already humming quietly, and both of them carrying a familiar name – International Harvester! No Kidding!

Those tough corn-binders were still humming when we sold the farm 22 years later. Frank tells me that the fridge latch eventually wore out and had to be replaced – $7.50. I have to think that Mom was lucky that John Deere didn't make appliances. She liked bright yellows in her kitchen, but two green monsters would not have turned her crank.

Along with the electricity, of course, came a water carriage system. Our well was 200 feet north of the house – the barn 300 feet south, and proper backhoes were far too expensive. We started to dig the 6 ½' deep trench by hand and after 20 feet or so realized we would never beat freeze-up – maybe two freeze-ups – so plan B clicked in. Mr. Wood in Fort Frances had a hoe which sat on a frame with two truck tires. It could be hitched to the B drawbar and the tractor hydraulics were strong enough to run it. The operator (Walter) had a seat at the controls, and though he had to change to the tractor seat to move it, things went fine and dandy.

While the trenching was underway the barn was being prepared. Cast-iron water bowls were installed, one between each pair of cows and were plumbed to the point where the water line would enter the barn. When the trenching was done the plastic main led from the well to the house north basement wall and connected to a pump and storage tank and continued out the south wall to the barn. The pump was primed, the switch was thrown and the cows had instant unlimited water – Wowee!

Now here's the part only farm folks will understand: The water came in one side of the basement and out the other – right? The cows had water to burn and what did Mom get? One tap! This is not a misprint – she had one cold water tap in the basement!

Mom understood – the budget had taken a whack, the cows brought in the cash, and with water at hand 24/7, milk production increased. Mom would have a sink upstairs with cold and hot water next spring.

Sidebar: The house was always last on the list. New machinery and what-not kept bumping the indoor bathroom back and we had to use the little brown shack for a few years yet. Last year's catalog still sat out there, losing pages until it was replaced by this year's last year's. For some reason the ladies' lingerie pages were always the last to go – them were the good old days.

Other Power Benefits

We now had an electric toaster – the most marvellous of modern inventions. We still ate porridge for breakfast but fresh toast made a tasty side-dish. A bed-time snack of freshly buttered, toasted homemade bread hit the spot and because you made it yourself, you could slap on the butter.

-The cows no longer needed to walk down to the well, so were kept in on cold days. On warmer days they were let out for exercise and at first, driven by habit, they would wander down to the water trough and stand around, figuratively scratching their heads, wondering why they were there. (I can dig it – I do the same thing ten times a day, now.)

-The horses and young stock still drank at the well, but the following spring the water trough was moved to a spot beside the barn door and when needed was filled using a hose. At the well, the big hand pump sat alone and unused for years. The wooden platform decayed and the pump now sat on the steel casing. Once in the odd while in summer I would give the handle a stroke or two, and rusty water flowed, soon turning crystal clear. "Thanks, buddy," said the pump.

-The lantern and coal-oil lamps were put away, but not on a back shelf – we were as yet unsure of Ont. Hydro's reliability. Eventually they gathered dust and the smell of kerosene faded. It was not an obnoxious aroma – we had lived with it for many years.

-Mom could now cook more efficiently, shop more efficiently and leftover food was un-wasted. Also, with a door on the fridge and a lid on the freezer, she could scam Dad ever so gently. Dad had two food rules – shop locally and don't ever buy margarine. Now when Mom went to town on a Saturday, usually with a son or daughter driving, she could visit the Shop-Easy and Safeway. She would pick up veggies and such, stuff that was just different. She knew margarine was cheaper – she used it for baking and it went into the freezer at home. (Did Dad ever catch on?)

Frank's Sidebar: In those days, pop-up toasters were a premium item – well past the budgetary restraints of a farm family. The less expensive Eatonia model we had was manually operated. You put the toast in, turned the lever to lower it which turned on the heat, then rotated the handle back up when you thought the toast was done – no easy feat at first.

Our nieces from Urban Utopia were visiting, and being used to the most modern of appliances, one of them (we never quite got them to fess up which one) put in a piece of toast, turned the lever and walked away to play. The untended utensil burned the toast to ashes, and was not discovered until the paint on the bottom of the cupboard above it started to burn. Mom ran in, grabbed the ever-present jug of lemonade from the fridge and quickly doused the flames before they spread. We almost lost the house that day.

Power led to a physics lesson. One Saturday afternoon I was messing around in the front yard with a five-foot piece of '3/4" rope, knotted at both ends. Where I had found it and why I was messing with it is lost in the back pages of memory, but I think I was waiting for the two pains-in-the butt to impress them with my rope-tossing skills and I spied the hydro line. I wondered if I could throw the rope clear over the lines, which I did, except on the third pass I didn't – the rope draped over the two conductors. Uh-oh!

I did a 360-degree head check – good – nobody around. Now I considered my options, which didn't take long, because I had none. A length of binder twine with a rock tied on the end might work, but adults had already clued us kids in. Pre-power we had flown homemade kites in the fall, and kite flying in the vicinity of the line was dangerous. I decided to leave the rope – there was no smoke or sparks – what could happen?

What happened was that a few days later it rained, the lines shorted out and tossed the breaker at the transformer. Hydro was called, the bucket truck fixed the problem and when they found a dumb kid was at fault they laughed and didn't bill Dad for the service call. (I had tried the "Who, Me?" defence with no success.)

But I learned that water will conduct electricity.

Other Memories of Crozier – 1950-55

A **Mid-Winter Rescue – Super Heroes Never Sleep**
Important info: take notes.
Time: Mid to late February 1955, Sunday, 2 am.

Location #1: Short's Corner – a four-way intersection where Ont. #611 now intersects with Pihulak Road heading east and Domanski Road heading west. (I'll bet few people remember that it used to be Short's Corner on the old gravelled highway.)

Location #2: The four-way intersection ½ mile south of our farm house. Domanski Road essentially ends here – the road that continues west is unplowed and is a packed sleigh trail where we haul pulp from our wood lot. Along the north side of the sleigh trail is a 50 cord pile of eight-foot spruce. Along the south side is 2 ½ feet of snow – luckily, we have just started to cut balsam.

Location #3: Napoleon Meilleurs' house near the Roddick school on #611. There was a party on. Nap played a mean fiddle and Walter could strum up a chord or two. "Play guitar, fill fruit jar and be gay-o."

Players: My brother Walter and Ross Peterson. Walter had taught school at Roddick last year – this year he is teaching at McCrossan/Tovell in Bergland. Ross Peterson's parents own the Bergland Store and he and Walter became close friends, remaining so until they cashed in a few years ago.

Bench Strength: Me – sound asleep, warm and comfy on a -40 Fahrenheit winter night.

Equipment list: Peterson's brand spanking new, fully-loaded green-and-white 1955 Ford Ranch Wagon. One good as new green and-yellow John Deere B. One better than new Percheron cross team.

The Incident Unfolds:

The party at Roddick wraps up at 1:30 am and our two loose-as-gooses guys head north. Ross wants to go to Fort Frances and across the river to the Falls to continue to party. Walter is partied out and wants to go home to bed.

At Short's Corner, Ross asks "Which way?" and Walter says, "Turn left." By now Ross can't tell east from west, so he follows orders.

Walter dozes off, pops awake almost two miles later, says, "Corner ahead!" and it's already too late. The Ranch Wagon hits the snow and it's a good thing Ross doesn't even attempt to turn right. The packed sleigh trail eases the car off to the left, missing the pulp pile and plowing its own road a hundred and fifty feet onward – and at 2 am Walter wakes me up.

He wants to harness the team, and I want to take the B. He is right, I am wrong, but I win the coin toss. The B always starts – that's a given. But it has a five-gallon cooling system and we never put antifreeze in John, because in the winter-time he is only used every second or third week to run the hammer mill. So I carry a pail of hot water from the house, pour it in the tractor and start it, adding cold water from the barn. Now we both ride the B to the corner trailing a 20-foot logging chain.

Of course the B won't budge the car – it can't find traction in the deep, loose snow, and we don't try to jerk the chain, being unwilling to pull the wagon out piece-by-piece.

I go back for the doggone horses, leaving Walter and Ross to shovel snow so the chain can be hooked to the rear axle.

Back at the ranch I harness the team before draining the tractor. (By now the B has cooled enough so that I can unscrew the drain plug without burning my fingers.) Now, with the double-tree sliding on the packed side road I walk behind, and it's a slow trip – the horses are not yet up to operating temp and they wonder why they are out here in the cold moonlight. At the accident scene the chain is put on the axle, and with a lunge and two farts, Doll and Bess walk the Ranch Wagon to high ground. Off the boys go to the house, leaving me to trot along behind two horses eager to get home – the exercise warms all three of us.

Walter waits at the barn and puts the team away. Ross has already crashed in the house. The Bat Signal fades in the clear, cold sky, I hang my cape on a hook, and goose feathers never felt so good.

(No animals or machinery were hurt, but a hundred pounds of snow had to be dug out from the Ford's engine compartment before Ross motored home to Bergland.)

The Annual Chimney Fire:

This actually was an annual event. The date was flexible – any time in January – but the time was always midnight to 2am. It would start slowly, picking up speed until we could hear crackles and rumbles as hot-air-driven flames cleared the built-up creosote from the ceramic-lined chimney. I was always worried, as I'm sure Judy and Frank were – we followed the lead of the adults. (Years later a light bulb finally glows – of course they wouldn't tell us not to worry – we had to be alert for a possible evacuation.)

But the fire stayed in-house, so to speak, Dad or someone would stand outside with an eye on the roof. There would be no firefighting attempts if the roof caught fire. The house was two full stories high with a 1/1 pitch roof.

But Mother Nature had a plan. She helped start the fire by dropping to 50 below so the wood furnace was working hard that evening. She then fought the fire at the same time. It takes more than the odd creosote ember to entice a snow-covered asphalt shingle to burn at that ambient temperature.

And one fire was cleaned the pipes for another year – sleep tight and worry not.

Except – we lived only ¼ mile from the CN line and there was often a night train. One came by a few nights after our first Crozier Creosote Cleaning and maybe the whistle woke me, but I can only remember hearing the rumble, which sounded exactly like a chimney fire. I could only stand it so long and I got up to feel the chimney, which was cool, and when the rumble faded away, so did my concern.

Sidebar: George Hyatt returned from tank-driving duties in Korea and worked for CN in Fort Frances until he went farming full-time. Many years later George told me that when they made up a mixed train 35 cars was the max. Now, the head end units are pulling into Rainy River as the tail end car leaves Fort Frances.

Table Scrap: Customer Service and Bill Collecting – '50s Style.

George Hyatt lived five miles north of Devlin and he needed a new chain saw. Fred Warren lived in Gameland, fifteen miles from Rainy River, and he sold chain saws. George called Fred and Fred had the model George wanted in stock.

"Save it for me," says George, "I'll pick it up in a couple of weeks when I have the cash."

"Tell you what." says Fred, "I'll put it on the train and you'll have it tomorrow. Mail me a check when you can."

That was Freddy for ya. Everyone has a similar Fred Warren story to tell, but Freddy Warrens are a rare find these days.

Rockets, Trains and Soft Summer Nights.

Living that close to the tracks was kind of cool. We crossed them twice daily on our way to and from school, and if we were lucky a train would go by and we'd get a wave from the engineer or head brakeman. We'd also get one from the caboose if the tail-brakey wasn't having a snooze – he might not be needed until Emo or Fort Frances.

The section gang used "rockets" to warn the engineer that track work was ahead. There were always two placed a short distance apart. The first time I heard them I was sure someone was shooting at the train – it sounded like two shotgun blasts. One day we took a track detour to see if we could find an expended rocket shell – we never did.

On warm summer evenings, we'd open the sashed, screened windows upstairs and drift off to sleep listening to whip-poor-wills and far-off coyote or wolf howls. A train might go by and the lonesome whistle call never interrupted the birds, but if a family of foxes were yapping, they would stop until the train faded away.

One evening I was about to drop off when I heard a strange sound, a loud one, as if someone was playing a giant accordion. It was a diesel horn – the first one I had ever heard! It took a long time before we saw one on the tracks – I think they only ran them undercover of night to befuddle kids.

When I finally saw one, it was slightly disappointing. I expected a red, white and star-spangled blue deal as seen in the Saturday Evening Post. These were CNR orange and black, but they were shiny and new, and I decided they were OK.

Of course we did the penny on the track thing – not too often – pennies were hard to come by, but we got a wafer or two. I know it seems unbelievably ridiculous, but we sort of hid, hoping the engineer wouldn't stop to chew us out for almost causing a derailment. What a silly, unsophisticated group we were – I'm surprised we even learned to tie our own shoe laces.

I finally got my genuine "Red Ryder" BB gun. I had lobbied many years for that prize, but by the time my whining bore fruit I was already too old for it. I wasn't allowed to rub out Judy or Frank (unfortunately) – I could only "plink" stuff. I wouldn't shoot birds, dogs or cats and I couldn't hit anything I aimed at anyway. Heck – I could only hit the barn if I stood close enough.

I preferred to plink with the Winchester single-shot .22 ("shorts" only.) The .22 shells came in boxes of fifty. "Long rifles" cost $1.50 per box, "longs" $1.00 and "shorts" were $.50. Shorts couldn't do much damage other than break a glass bottle, but glass was out. Dad didn't need broken glass to deal with, so I plinked tin cans.

Empty oil cans were the best – Veedol, White Rose or Quaker State. If you filled them with water they sat still while you drilled the centre of "Os" and "Qs". The best test of marksmanship was an empty oil can. If you caught it just right near the edge the .22 short slugs would follow the inside curve, leaving a groove before falling spent to the can's floor.

Table Scrap: A True "Short" Story.

A few years ago in (place name censored) a local drug deal was going down. The buyer and seller obviously did not agree, because as the buyer walked back to his car the seller popped a cap into the back of the buyer's noggin. Man down!

When the paramedics arrived the guy was already standing up, complaining about a headache. At the hospital they found the .22 short had barely dented his skull – three stitches and an aspirin.

Chalk it up to old ammo and a thick head – a lucky break for both shooter and shootee.

Sidebar: At the height of the BSE scare I saw a photo taken in England. A large pit had been dug and helpless cattle herded in. British Army sharpshooters were called in to shoot fish in a barrel. Kind of shakes one's faith in regular army riflemen, doesn't it?

B erry picking was fun, as it always has been and always will be. We picked blueberries now. There was the odd Saskatoon bush amongst the poplar trees, but here in Ontario the birds got to them first. The best thing about blueberry picking were the pleasant smells – spruce, pine, balsam, Labrador tea, moss with sweet-smelling water not far beneath, and even on the hottest days it was cooler in the blueberry bog. Later on a neighbour tipped us off about low bush cranberries – easy to find in September. Home-made cranberry jelly beats store-bought hands down. Wild strawberries were harder to find, as were wild raspberries.

Wild plums were new to us, but most when ripe were too bitter to eat off the branch. If you found a grove of sweet plums, you kept that a secret to yourself – they were too tasty to go to the canning jar. Chokecherries abounded as they do everywhere. Kids will eat anything semi-tasty, but a handful of chokecherries puckered our cheeks for hours.

Gooseberries were a rare treat. They were green when ripe and thus hard to spot, and the time between ripe and when the spiders found them was short.

How about puff balls? By September the fungus turned into brown grass apples, and when hopped on gave a satisfying "pop" with exploding brown smoke. I guess we were junior environmentalists – helping Mother Nature spread spores.

In the spring of '52 Mom made a batch of dandelion wine! Many gallons of dandelion wine arc (and were) made every year by many people, but when Mom made the announcement I was shocked because alcohol was, and always had been, absent from our house. While our parents never thumped the sobriety drum, it was just a mutually respected unwritten rule.

It takes a wagon-load of dandelion heads to make the wine and it was up to Judy, Frank and I to pick them. We started in the front yard, then expanded our hunt to the pasture and hayfields. Bushel baskets were filled and refilled and by the time our contract was filled we smelled like dandelions and our fingers were stained with dandelion milk. We didn't have any bumble bees around that year – they had gone to yellower pastures.

The project yielded two bottles of wine – I figured about a 1000/1 volume reduction. The bottles were tasted, rated drinkable and went into the fridge where they sat unopened for months.

That fall a couple of Rod's high school friends visited and the wine came out, glasses were poured and everyone complemented Mom on a job well done. We rugrats got no recognition nor did we share in the wine-tasting ceremony.

(In the '70s, before Dad died, a mickey of brandy was always in the International Harvester fridge for medicinal purposes – the only liquor that was ever in our house other than the wine batch.)

The Saturday Evening Post was always part of my early years. Later in life I realized the Post was a right-wing-slanted publication, so Dad probably voted conservative. I was totally disinterested in politics but there was enough stuff in there to keep me busy for seven days – I read it all.

First I would scan it front to back. I liked the full-colour advertisements. From 1949 to '51, "There's a Ford in your Future." (including crystal ball) or "When Better Cars are Built, Buick will Build Them." There were '51 –'53 "Twin H-Power" Hudsons, fresh off another Nascar win. Lincoln and Cadillac ads were simple, straight-forward and without much hyperbole – usually a couple in evening clothes stepping out at a fancy restaurant or hotel – pretty boring. Other manufacturers showed families having fun on the road and convertibles always had tops and windows down. The roads were smooth and dust-free – the photo shoots must have been done in sunny California. The only trucks I can recall were Studebaker stake jobs – always on the farm. Pickups were notable only by their absence – I guess Post readers didn't drive lowly pickups.

Other ads caught my eye – Exide batteries with "Power to Spare" or Ray-O-Vac flashlight batteries (Steel Clad) and every issue had a small ad showing a kid in his pyjamas, rubbing his eyes, and with a tire on his shoulder – "Time to Retire with Fisk."

But the best ad of all appeared only once a year in early January. It was either Atlas Tires or Mobil Oil and in full vibrant colour, it showed every licence plate north of the Rio Grande including the Northwest Territories Polar Bear cut-out. (NWT must have the longevity record for licence plate shapes.) With that ad as my guide I would be the licence plate recognizer champ every spring.

Heavyweight title boxing matches were broadcast, sometimes two in a 12-month period. Dad liked them, so we got to listen in. Rod said Rocky Marciano was the greatest, retiring undefeated, but I thought he only faced bums, avoiding the best of the rest. My favourites were Ezzard Charles and Sugar Ray Robinson. They were two light-heavies who weren't afraid to tackle heavyweights, and I always pulled for the underdog.

Table Scrap

In September of '63 I had stopped off, en route from Yellowknife to Sudbury, to spend my accumulated time off at the farm in Crozier. Oldest brother Walter was teaching and living at Stratton, so we planned a bit of a get-together. Sonny Liston was fighting Floyd Patterson and the plan was to have a beer or two while listening to the fight on the radio.

I arrived with a dozen Blue, the pre-fight ceremonies were almost over and we opened the case.

DING! – Round One – and we didn't have a bottle opener. I went to the kitchen and sister-in-law Joyce poked around in a drawer, saying, "I know there's one in here somewhere."

I walked back into the living room, just as the ref counted "Nine, ten!" It was over at 2:09 of the first round, and I don't know who was more disappointed. Floyd's career was done, and so was our mini-party!

Table Scrap: How Times Change

Actually, TV had been available in our valley for years. Some "rich" people could afford to erect a tall aerial and were willing to peer through the "snow." The nearest transmission tower was at least 100 miles south and thus on many days the American stations were lost in a blizzard.

But it came to pass that our government finally decided to bring us CBC – TV. For years microwave towers had been feeding our neighbours along the Trans-Canada and now were being built southward from Kenora. They probably thought we Valley hicks were in danger of becoming "Americanized."

The winter of '62-'63 interrupted tower construction before the "fenceline" was finished, and Nestor Falls had clear-as-clear-could-be CBC-TV. The Rainy River Valley had to be patient – the signal would reach us in the summertime.

In late April, once again on seasonal time off, I was visiting at the farm. The Stanley Cup Finals (Toronto/Detroit) were on the tube, so I picked up Walter at Stratton and off to Nestor Falls we went.

We watched the game in the bar at the Nestor Falls Hotel – packed house that night – how enjoyable. In those days Esso was the sole sponsor and the game was never interrupted by commercials. We only had to put up with Murray Westgate while he pretended to pump gas between periods.

We drove sixty miles to watch a hockey game! Now I keep my remote at hand – Sesame Street is more interesting.

Table Scrap: How Times Change II

With TV available by summer '63, we all chipped in and bought Mom and Dad a set for the living room. Had they ever watched a TV before? I doubt it.

They soon got into the swing of things. Dad could watch the news (although he still sat by the radio at noon for the Farm Report.)

Mom really enjoyed the new experience, but it took her a while to catch up to the real world.

One of their favourites was "What's My Line" – Dorothy Kilgallan and Anne Francis were the two ladies on the stage set. Every week as the panelists were being introduced, and as Dorothy and Anne walked into view, Mom would clasp her hands – she was <u>always</u> amazed. "Another new dress this week?!?"

You have to understand – all her life she had been on the "One New Dress Per Decade" plan.

So, by and large, the fifties were the good old days. World problems were not front and centre, and TV wasn't around to tell us what we didn't need but had to have. Problems were usually your own fault (I was sure about that) but I know now that my problems were peanuts compared to some other kids. Maybe that's why I never wanted to grow up.

Chapter VI – New Horizons

I started High School in 1953 and my world started to open up. I now had 749 fellow students. Wow! In Manitoba we were lucky if more than ten kids could be chased out of the bush, and at the one-room deal a mile from the new farm the total enrollment never exceeded thirty – but now – 750!!

Adjusting to the rural/urban student body mix was difficult. Most of my fellow country hicks were strangers and the multitude of pretty girls scared the dickens out of me. I hunkered down and tried to stay as anonymous as possible – pretty hard to do when you're a foot taller than anyone in your age group. And then in 1955, I turned sixteen and got my drivers licence! What a game changer – fiddle foot training was about to accelerate.

I had been told that I was born shortly after midnight, so I was sixteen years and nine hours old when I walked into the licence issuer's office that morning. It was a school day, but I didn't give a rat's behind – I was after that little piece of green paper – my ticket to freedom!!

Bob Trenchard was the licence guy, and other boys had told me that he could be tough to deal with. A driving test would be necessary – both hands on the wheel and I'd better be careful. Forget about parallel parking, that was only necessary if I wanted a chauffeur's licence – so skip the parking bit. (In Ontario back then there were two licence classes – operator and chauffeur. The first class allowed you to drive a car or pickup. The second gave you the right to drive anything, including 18-wheelers.)

So forget it – I figured I'd never be issued the chauffeur deal, but here's how it went down.

Me: "I'd like to get my drivers licence, sir."

Mr. Trenchard: (After studying me very closely, making me very nervous,) "You're a farm boy, aren't you?" (What tipped him off? Was it because I tucked my vee-neck sweater into my jeans?)

Me: "Yes, sir."

Mr. Trenchard: "You've already been driving for three or four years. In fact, you've probably driven your dad's pulp truck – am I right?"

Me: "Yes sir," (extremely nervous now. Where is this going? Am I about to be reported to the cops?)

Mr. Trenchard: "I don't test farm boys and farm boys always get a chauffeur's licence." – and I walked out of there feeling like I had a million bucks in my wallet.

Maybe if I had stuck to shank's mare things might have played out differently. I might have become a "normal, upstanding citizen" –a "decent sort of guy" – a career in the pulp mill – a teacher – or perhaps a suit in my future. But nope – I had a driver's licence and the open road beckoned. How little did I know that a few years onward the pavement would end at the edge of the northern boreal forest.

Up to now I had been an indifferent student. With a licence to drive I now became a different student, you might say. I was smart enough to maintain some sort of a grade average without cracking a book, but I was much better at getting kicked out of school than actually staying there. I found it was tough to string a full ten months of grade 11 together. I was becoming a rebel (junior grade) and all rebels must go through basic training. And I became a car guy – want to hear some car guy stories?

The Devil and the Guardian Angel – Which One Will Win?

It was the homecoming game for the high school football team – a fine October indian summer Saturday in 1955. The Muskies played home games in International Falls. They had a dandy football semi-stadium and we didn't. It was a trade-off – Falls kids played hockey in our Memorial Arena.

The game was preceded by the Homecoming Parade. Some businesses (i.e. Green's Furniture) would let a responsible guy use their company truck and other kids used Dad's stuff. The vehicles were decorated, loaded with cool guys and pretty girls and were led by the Homecoming Queen and King sitting on the top boot of a shiny convertible.

The parade was off to the game, with horns tooting and thousands lining the parade route. Down Scott Street it snaked across the International Bridge (no toll or customs) and on to the stadium. In 1954 and '55 the royalty rode on J. A. Mathieu's 1953 blue-on-blue Buick Riviera – a limited production unit. I think the only road miles that Buick convertible saw were parade miles.

This year I had my licence to drive, and I would drive the family's '50 Meteor in the parade.

The plan was to motor in to the Fort, pickup four buddies and we would decorate the car with help from a chick or two (we hoped.) Like all good plans, this one had a flaw – and the flaw was my eleven-year old kid brother. It was a "you scratch my back and I'll scratch yours" deal. If I wanted to use the Meteor I had to deliver him to his violin lesson, collect him a half-hour later and return him to the farm. Bummer – it seemed workable, but I knew we would not be trimming the car very fancy-like. The lesson was over at 6:30, the parade started at seven, and although it could easily be done, the decorations would not last very long at 60 mph. I dropped Frank off and picked up the first three friends. We all agreed there would be time after the lesson to slap on a few strategically-placed crepe paper roses, and join the tail end before the parade crossed the bridge.

When we got to the fourth guy's place (we'll call him Ed) plan B took shape immediately, and what a plan it was. His Dad had bought a new Super-88 Oldsmobile last year and in a moment of weakness told Ed he could put that beauty in the parade. Dad must have had an enormous brain fart, because, as Ed did not yet have a licence, I was given the keys!!! Quickly, before Dad had a chance to change his mind, we swapped cars.

Now what? I still had to pick up kid brother. I thought they should trim the Olds while I delivered Frank home with the Meteor and they would have Ed's car ready for parade duty when I returned.

I lost the vote by a landslide. The yellow and green Olds needed no further beautification, so the majority decided we would all take Frank home in big car comfort.

What a dream to drive! She was smooth, silent, and beneath the gas pedal, overhead valve horses were eager to run. Someone said, "Let's see what she will do," and that's all it took. When we rounded the wide bend east of Hampton's Store we were up to 85 and little brother (little shit disturber) who was sitting between me and Ed was practically bouncing in his seat. "Faster!" he said, clutching his violin case – "Faster! Faster!" echoed by the other boys.

It was dark now, but the Olds had good headlights, and ahead was a set of tail lights. We were nearing the crest of the rise at the Crozier school-house, and I had no business passing on a blind hill, but I was keeping an eye on the hydro lines – oncoming lights would reflect off the wires.

We topped the hill at 105 on the wrong side of the road and two dim as dim could be headlights were almost upon us. I hit the power brakes and yanked her right, but the car we were overtaking was dawdling – no way were we going to snub up in time to avoid a rear-ender. We missed his back bumper with a foot or so to spare and already I had let up on the brake and was cranking left.

The ditch was smooth with a longish, low slope and the Olds was heavy, with a low centre of gravity – a perfect combination to save our stupid asses. The Olds slid sideways perpendicular to the highway with our headlights blinding the wide-eyed lady in the passenger seat of a '51 Meteor. It was obvious that she was scared enough for all six of us – we were too wired to be frightened.

I momentarily toyed with completing the sideways pass, but I knew Miller's driveway was coming up, so I let the Meteor go ahead before converting the slide to a gentle climb back onto solid pavement. We should have turned around and hid, but the other guy wanted to stop, so we passed him legally and turned south on our side road.

The trouble was – we knew the car, we knew the owner knew Ed's dad, and we hoped he wouldn't recognize the Olds. Fat chance – how many bright yellow-and-green cars were around – but we hoped he wouldn't rat us out.

We were a pretty quiet bunch now. We dropped Frank off, sworn to silence – a promise which I believe he kept, and quietly returned to town on a back road. We scanned the Olds – it looked OK, and we left it at Ed's. We never did hit the parade. We drove around for a bit in our Meteor, completely drained, and went home early.

What we hadn't seen in the dark and what we couldn't have fixed up anyway were the hunks of grassy sod on the Olds' back bumper and the tufts of grass on the right rear wheel between the rim and the tire bead – we were that close to peeling off the tire and rolling over.

A simple summation: The Devil made us do it, but our Guardian Angels saved us.

(Ed's dad was a millwright. Eleven years later I would be working in the mill and would run into him from time-to-time. I never could look that man in the eye.)

Hot Rod Lincoln by Charley Ryan and the Livingston Bros. (circa 1955)

Have you heard the story of the hot rod race that fatal day
Where the Ford and the Mercury went out to play
Well this is the inside story and I'm here to say
I was the greaser that was driving that model A.

It's got a Lincoln motor and its really souped up
And a model A body makes it look like a pup
It's got twelve cylinders and it uses them all
With an overdrive that just won't stall

It's got a four barrel carb, and dual exhaust
With 4:11 gears you can really get lost.
It's got safety tubes and I'm not scared
The brakes are good and the tires are fair

I left San Pedro late one night
The moon and the stars were shining bright
Everything went fine up the grapevine Hill
We was passing cars like they was standing still

When all of a sudden like a flick of an eye
A Cadillac sedan just passed us by
A remark was made "That's the car for me"
'Cause by then the tail lights were all you could see

Well. the fellas ribbed me for being behind
So I started to make that Lincoln unwind
I took my foot of the gas and man alive
I shoved it down into over drive

Well I wound it up to 110
I twisted the speedometer cable off at the end
I had my foot feed clear to the floor
I said, "That's all there is, and there ain't no more."

I went around a corner and I passed a truck
I whispered a prayer just for luck
My fenders was clicking the guard rail posts
The guys beside me were white as a ghosts

I guess they thought that I'd lost my sense
The telephone poles looked like a picket fence
They said "Slow down, I see spots!"
The lines on the road just looked like dots

Smoke was a-rollin' out of the back
When I started to gain on that Cadillac
I knew I could catch him and hoped I could pass
But when I did I'd be short on gas

I went around a corner with the tires on their side
You could feel the tension, man what a ride
I said "Hold on, I've got a license to fly"
And that Cadillac pulled over and let me go by

When all of the sudden the rods started knockin'
And down in the diff she started a-rockin'
I looked in my mirror, a red light was blinkin'
The cops was after my Hot Rod Lincoln

Well, they arrested me and put me in jail
I called my pop to go my bail
"N" he said "Son, you're going to drive me to drinkin'
If you don't stop driving that Hot Rod Lincoln"

In 1956 Dad bought me a car, and what a car it was – a '51 Monarch convertible. (The Monarch was a Canadian car based on the Mercury with a few modifications of the shiny parts.) Our local Ford/Monarch dealer drove a new convertible every year, selling it when the next models came into his showroom.

A local guy whose grand-dad had left him $10,000 bought the '51 Monarch and wore out the ten grand and the car in less than 2 years. I had seen it many times as I walked to school in grade seven and eight. It would cruise by with the top down, full of pretty girls and happy guys – and it sure looked like fun! After the money and the car were used up the convertible sat, lonely and dejected in the back corner of a car lot until my dad bought it for me – eighty dollars for that treasure. (Incidentally – less than 400 Monarch ragtops came off the assembly line in 1951.)

It had power seats, power windows and a power top – and not one of them worked. The top could be pushed down by hand but the windows stayed up, refusing to budge. It had brown leather upholstery and you could tell the seats were well-padded because all the stitching was gone. The hood was held down with haywire and it had no low gear.

I filled the car up with five other guys one night and we went to a drive-in movie on the American side. With the top up and two guys in the boot we only had to pay admission for four. We were a little concerned that someone might see the two stowaways crawl out when we parked at the microphone stanchion, but we needn't have worried. By the time the Monarch climbed the insignificant grade to the mike pedestal we were enveloped in blue smoke – she burned a tad of oil. Behind us horns were blowing and people were shouting, "Turn that thing off!" It rained after the show started and the windshield wipers didn't work so we went home and never did see that movie.

All through the next month that convertible was busy and school played second fiddle. We had fun, fun, fun 'til Daddy took the Monarch away. He sold it for $120 – not a bad deal, car-wise.

In June of 1956 I got kicked out of school, and it was a late harvest of a seed that had been planted two-and-a half years before.

While it may seem unfair to blame my unsuccessful high school career on one specific teacher, this one certainly deserves to be cast as the villain – he fit the part and played it well.

I locked horns with him on my first day of high school in 1953. My parents, with dreams of a future academic in the family, had signed me up for French and Latin – Latin, for crying out loud! When I saw that on my list of classes, I boogied straight to the office and swapped it for Music, figuring (I figured right) that Music could be aced without cracking a book.

At noon, less than three hours later, I was standing at my locker when Mr. F (and this will be the last time he will be called "Mr.") cornered me. He was irate, and he demanded an explanation – why had I dropped <u>his</u> Latin class?

I was fresh off the rural turnip truck and he scared me, yet I had to wonder – did he own the Latin franchise? I mumbled something and walked away, but I could feel those mean eyes drilling into the back of my skull. What I didn't know, but would later find out, was that he was already plotting revenge.

Throughout Grade Nine and Ten, whenever I met him in the hallway, I got the evil eye and I learned to avoid him if I saw him first. (I realize now that F was a functional psychopath and psychos, like elephants, never forget.)

So in my first year in grade eleven, I was doing barely OK in school, but I had a good chance of passing. Then two days before final exams, I skipped a class, F caught me, and had me kicked out of school. He had beetled down to the office, did a forensic investigation, pointed out that I was essentially a part-time student, and I was OUT! I was not even allowed to write my finals!

I was aghast! Most of my other teachers were also aghast! My geometry teacher was very apologetic that he could only give me a final mark of 50% (I had written two mid-terms at 100%.) And, my parents were very aghast! I had a lot of 'splainin' to do – mainy 'splainin' why I should not be shot at sunrise. So although F might have thought he had won the war, he had only won a battle. I would return to fight another day.

In Sept. I hit Grade Eleven again, determined to do better, and Holy Hades! – I would be taking French with F! French had never been a problem for me, mainly because up to now I was taught by Miss D. – a stern, but fair lady – a good teacher. I had to see how this played out.

The first day in his class I saw an opportunity and I took it. The classroom was only ¾ full. He had a dozen or so students taking French and the others were in a "study" mode. They were not taking a full schedule – they were taking some subjects to enhance marks, and thus sat at the back of the room. I joined them at the rear, keeping an eye on F, who seemed to be ignoring me. I waited for him to scan his class enrollment and centre me out, but he never caught on. I had him pegged now – he was a lazy psycho – a dumb one, too, and I had him by the short hairs. I sat there until Christmas, hiding my open French text book in my 3-ring binder. (Guys sitting near me thought I was nuts – they were hiding their Playboys.) I "audited" the course, mostly auditing his teaching methods, which mainly consisted of keying on his poorer students. He didn't help them at all. He put them down, continually scoffing at their questions until they gave up, and it really pissed me off. How was this man ever allowed to teach anyone??

"When I think of all the crap I learned in high school,
  It's a wonder I can think at all," – Neil Simon."

Christmas exam time rolled around. Only those writing the exam were allowed in the room, and I waited until the last one had entered before I went in. F, of course, jumped up and told me to leave. I told him I was writing the exam, suggesting he check his class enrollment, which he belatedly did, and it was pretty to watch.

His face turned white, to red, to blue, and when he opened his mouth, he couldn't speak. He had no recourse but to allow me to write, and although he red-lined every conceivable ditty-bum error, he had to pass me. It must have broken his little psychopathic heart.

I had won this battle, but I knew I could never win this war if I stayed on, so I won by default. I quit school.

Sidebar: Six and half years later I was spending a few days at Copper Cliff tuning up for an Inco summer in NWT. I was heading to my car in the field office parking lot and walking down the sidewalk was F – I couldn't believe it! I figured bygones were bygones, so I sauntered over, calling his name. He turned, and without a greeting, he said, "I remember you. You're Bob Durnin. See? I never forget a face."

What an egotistical prick! I turned on my heel and walked back to my car, wondering why I had gone to the trouble to play nice.

He's gone now, and if he reads this, it will have to be written in stone, because paper burns. I may yet get to meet him again, face -to-face – and I can hardly wait!

In Sept. 57 I went back to school for a third crack at grade 11. This time I decided to be independent so I got a job pumping gas for Jim Witherspoon at his gas bar on the west end of town. The idea was to work from midnight to 8 am, attend school between 9 am and 4 pm and sleep the other eight hours.

It started out well, but by week three I was having trouble staying awake in the more boring classes and when things were slow at the gas bar. To me sleeping on the job was a much bigger faux pas than catching a little shut-eye in history class. Jim was far more understanding than my teachers and detention time mounted exponentially. At the end of September I quit the job to salvage the school year and then at Christmas I left school anyway – I should have quit school first and kept the job.

Table Scrap

One night at the gas bar was veddy interesting. It was around 1:30 am and traffic was light. A '50 Merc two-door pulled in – two good old boys who had obviously stayed at the bar until last call. I didn't know them, but I knew who they were – loggers who cut and hauled their own pulpwood. They had a loose cannon rep and were not regular customers.

I filled them up and gave them an official customer loyalty card – my first mistake. They wanted to know the deal and I explained it. A spiffy glass case on the pump island held glassware – nice stuff. The card was marked around the edges in five-gallon increments totalling 50 gallons and a filled card could be turned in for things from the display case. For instance, one card got you a nice water glass or a dessert dish, and other things (like a dinner plate) required two or more full cards. Top of the line was a beautiful cut-glass water pitcher. It had been in the case since day one.

The boys wanted the top prize. Once again I tried to explain the rules but they wanted that pitcher, and they were starting to get owly, and I was getting nervous. I kept hoping a car would stop in to give me a chance to disengage but no customers appeared.

Finally, to avoid a possible thumping I caved in, gave them the pitcher, and already I was trying to figure out how to justify this to Jim.

I wanted to see the last of these boys so I talked up the Merc. I opined it could spit gravel leaving the gas bar. They took the bait, backed up to the end of the lot and peeled out.

The local from Atikokan was due in at 2 am and it was never on time unless you were late. Our neighbour, John DeGagne, was on his way into town to meet the train. He was in the wrong place at the wrong time.

I was at the kiosk door when I heard the crash. The loggers, exiting to the west with pedal to the metal, drifted across the centre-line and hit our neighbour head-on, totalling the '50 and with lots of damage to John's heavier '53 Merc. I called the cops and hustled down to the scene where the two yahoos were stumbling around dazed and confused. Another car had stopped and they were looking after John's broken arm.

The police arrived and confusion reigned – who was driving? Both drunks insisted they had the wheel. The cops went to brief the ambulance people and I heard the boys whispering. It turned out one guy had already lost his licence and he was telling the actual driver to shut up – one of them had to keep a steering ticket to drive the pulp truck. The police asked me if I knew which drunk was driving. I was clued in now, and although they had been jerks, I didn't want to see them without a means to make a living, so I claimed ignorance.

As the argument raged on I could see that glass pitcher sitting on the front seat. I walked around the car, quietly reached into the open side window, retrieved my prize and took it back to the display case – I was off the hook.

After Christmas I was out of school and on the outs with Dad, so I bummed a ride to Winnipeg, went to my sister's and started to scan the help wanted ads. The late '50s recession was building and it seemed few outfits were hiring.

An ad caught my eye – "Excellent opportunity in sales. Salary plus commission." I called the number. I was told to attend a "sales orientation" at an address on Portage Avenue in downtown Winnipeg. An evening orientation? – that alone should have tipped me off.

The meeting was in an empty store with a "for rent" sign in the window. There were fifteen or twenty guys and I think we supplied the heat because we all kept our winter coats on. There was a wide range of citizenship represented, from semi-hoods to a semi-professional. (The semi-pro left early, needing no orientation to tell him where the bullshit lay.) We sat on dusty stacking chairs and found out "what's up." We were to sell encyclopedias – Grolier's Encyclopedias..

First the guy showed us our sales kit – a half-decent looking cross between a suitcase and a regular briefcase. It promised more than it held. Inside, along with customer contracts was a fold-out mock-up of what your Grolier's would look like on the shelf. It looked like they were thick books – A-Z would need 20 linear feet of space.

Next came the sales spiel – how to get your foot in the door, how to pick your household member target – all fairly low-key, but in strict order. The guy told us to memorize the steps. He said that we could wing it if called for, but the method was tried and true.

Somewhere along here I got up the nerve to ask him about the salary plus commission deal. He said there was no salary and if anyone expected said salary he might as well leave – and the semi-pro walked out. I hung around – jobs were scarce, and what the heck – at 40% commission I would become a wealthy, wealthy man on one sale per day.

We were sent home with our brown pretend-leather satchels. Before we left, we were assigned our sales districts – I drew an older residential section on the south side of the Assiniboine River including Fleet Street. I would start there tomorrow evening – prime time was seven to nine pm when the man of the house was present.

The next evening rolled around and I was ready to sell! I had memorized the sales pitch and as my brother-in-law drove me to Fleet Street, I thought and planned, and when he dropped me off, I was wired for action. I figured I might cover a block or two. He would cruise Fleet at 9 pm to pick up the successful super-salesman.

It was the coldest night of the coldest January I can recall and I soon found out that I hadn't planned too well after all. My sister had told me to bundle up, but I knew better. I didn't want to show up on someone's doorstep looking like Nanook of the North, so I wore no hat nor overshoes and my leather loafers squeaked on the frozen sidewalk snow.

Many front windows were dark – did folks go to bed early here? I tried a few front doors on houses that showed a TV might be on inside, but got no response. Belatedly I clued myself in – supper was over, tea or coffee was on, and side doors were my best bet. Two doors later my knock was answered, I got my left foot in the door, and immediately ate my right foot.

I can only blame the weather for my brain cramp. It was so cold, and I was so cold that when a guy slightly taller than Mickey Rooney answered, I blurted, "Is your mommy home?"

Mickey said, "Come on in and meet my wife," and I shrunk to half size – what a dummy! But I was freezing and the house was warm and toasty, so I entered the lions' den.

He invited me into the living room where I met his wife – a tall beauty who smiled and said. "Hello." She probably knew what was coming, so she went back to the kitchen to wash the dinner dishes, leaving me to learn my fate at the hands of her husband. What followed was a far more enlightening tutorial than the one our sales manager had given us last night.

He asked me what I was selling. I told him I represented Groliers's Encyclopedia. (I had already espied their set of Britannica in a Britannica bookcase.) He wanted to hear about mine and when I said he was already encylopediaed up, he insisted I continue, so I did so.

After I was done, he proceeded to tear my product apart. He wasn't mean about it – merely pointing out that Grolier's was bullshit and I had to agree with him – I was selling an inferior product. He had his revenge for my "mommy" mistake, but I had learned a thing or two, so we ended up even-steven.

I hit Fleet Street again and it was still cold out there, I had an hour to kill, so I bit the bullet, continued on my way and two or three houses later I hit paydirt.

A nice younger gent answered my door knock. I told him I was a representative of the finest product known to mankind, and he welcomed me into the house. I was ushered into the clean and simply-furnished living room and was given the place of honour in a comfortable armchair. Before I had a chance to start, other members of the family gathered around the perimeter of the room – from ankle biters to Grandpa and Grandma. Kitchen chairs supplemented the sofa and I had the complete attention of everyone.

I automatically went into my BS sales mode, but I was keeping my eyes open and my brain in gear. Only the younger gent seemed to understand me, although the others, when I made eye contact, always smiled and nodded. It didn't take too long for me to get a handle on this deal. These were obviously a recently arrived family of immigrants, possibly after narrowly escaping KGB punishment following the Hungarian Revolution. Of course they were a good audience – they were all learning English and Canadian Culture. Had they ever seen a cyclo-salesman in their home country? More likely it would have been a comrade enforcing a long-term Pravda subscription – buy or die.

I asked myself, did they have any money? Could they maintain the easy payment plan? If so, I could easily have sold them the Diamond Edition, complete with guaranteed-to-collapse cyclo-case and fifteen years of annual updates – but I just couldn't do it. In mid-sales pitch I folded my tent and told the younger gent to hold off buying an encyclopedia set until the kids got further on in school, and I even wrote it out for him – Encyclopedia Britannica – the best one for you.

When I left that house I actually felt pretty good. I had already made up my mind to turn in my briefcase, but what did I have to worry about compared to these nice people? I was in my own country, with extensive family support and the secret police were not keeping an eye on me – at least not yet.

Actually I did have something to worry about – it was only eight-thirty, it was still bitterly cold, and my ride would not show until 9. Onward on Fleet Street I went, with me being definitely more fleet of foot by now. I passed six dark houses and at the next one showing signs of life I knocked. An older gent opened the door and without a "hello" I said, "I'll promise not to try to sell you anything if I can come in to warm up," and he opened the door wide.

What nice, friendly folks! Grandma put on the teapot and we had a nice visit. I asked him why, with the night being so miserably cold, was there no chimneys on Fleet Street emitting smoke? He said this section of Winnipeg was hot-water heated from a central coal-burning steam plant – how interesting! At 8:55 I thanked them for the warm-up and hit the sidewalk to flag down Leon. On the way home I thought about the evening and how, even though a job may be the pits, there is always something to be learned.

The very next day I went back to the farm determined to get a "real job," which I did, and in February of 1958 I went to work at a bank in Fort Frances – my last stab at respectability.

I was neither good nor bad as a banker – sort of mid-range. One upside was that I was working with pretty ladies and each and every one of them was a whole lot smarter than me.

I was no great whiz at numbers. I tended to transpose figures and as a teller I had to balance out every day after closing. I'd struggle for half an hour or so before one of the ladies would come along and in thirty seconds she had me balanced. Then the manager passed the word – let him learn to do it himself.

Now the pressure was on – no one could go home until the kid balanced. While I was scratching my head and other parts of my body the girls would hang out at the back, tapping fingers on tables and itching to go home to get spiffed up for their evening dates. I felt like a fool.

And another downside – I had to wear a tie and I always felt I was choking. Suits, white shirts and ties – I've always hated them. But while the snows fell and the winter winds blew I was snug and warm inside. Then spring arrived and conditions reversed. Now I was stuck in a stuffy old bank while out in the real world flowers were blooming, grass was growing, and spring showers were washing the streets and sidewalks. Outside my barred window people were walking by – girls in summer dresses, guys in short-sleeved shirts, and I couldn't even take my suit jacket off – rats and double-rats!

But banking was nine to five – long summer evenings and weekends were free – so I bought a car. Set back and relax – it's time for another car story.

We are back in Monarch Country. This one is of certifiable gem quality – a 1950 two-door, aka "Sports Sedan." A local guy, apprenticing as a mechanic had customized it, had been caught drinking and driving, and had to sell it to pay his fine. I got a pretty good deal.

Like the '51 convertible, the '50 Monarch was identical to its Mercury cousin with

the only difference being the grille and badging. Removing the vertical bars from the grille leaves the cross-bars for a unique look – the easiest part of the customizing process.

Now the real modifications start. The chrome disappears from the hood and the trunk lid and the holes are filled. The Monarch already has an in-car hood release and another, hidden in the rear window shelf releases the trunk lid.

Now – fill in small dents and dimples, sand and smooth and repeat until it's ready for six coats of Alaska white. Shorten '57 Mercury Turnpike Cruiser skirts and install same. Two-inch lowering blocks at the rear leaves three inches of tire visible beneath the cruiser skirts. Dual exhaust tips are near ground level at the rear bumper – be careful when exiting a driveway. Install two dummy tear-drop spotlights at each corner of the windshield. Add Blue Dot taillights and '53 Olds Fiesta flipper hubcaps to the front wheels.

Now, with one hand on the steering wheel, your left arm resting on the open windowsill, Elvis on the radio, and dual Schmittys burbling at the rear – gas 'er up and chase pretty girls. I loved that car. It made James Dean's '49 Merc in "Rebel Without a Cause" look like Aunt Martha's go-to-church Plymouth.

We didn't booze it up much in our late teens, mainly because it was too expensive for our limited budgets. I mean really! Who would pay almost six bucks for a bootleg 12-pack when you could cruise main street non-stop for six hours on five bucks worth of gas? We did imbibe a bit just to show we were cool, but the group I hung out with were more sippers than chug-a-luggers. However, even occasional social drinkers have to be careful.

One memorable '51 Monarch convertible evening sticks in my mind. Six of us somehow scored a 12-pack – two apiece, if you are keeping score. As we drove around a bit we worried about having contraband in the car – what if we were stopped? I had a brain wave, and this will show you how smart I was. We stuffed the unopened bottles in between the top bars and the canvas top.

There were two beers left up there when I went home to bed. I got up in the morning and what to my wondering eyes should appear, but two unmistakable lumps in the Monarch's roof. We might as well have written "O'keefe Ale" on those two bumps.

Another Monarch evening – this time it was the white one. We went to the Falls and bought (legally) two six-packs of 3.2 and tossed them in the trunk. At Canadian customs, the officer asked me to open said trunk, but the handle on the rear window cable had been pulled off and I had not yet fixed it. I usually used a pair of vise grips to pull the cable, which I had in the car, but would not admit I had. I asked him if he had a pair of pliers, and he waved us on – close call.

On yet another evening in someone else's car we were once again doin' some cruisin' in the Falls. The car owner wanted to stop at the (censored) liqour store. They sold mickeys of pre-mixed vodka and he bought a bought a vodka/lime. We crossed back with the bottle in the glove box – pretty chancy, I thought, but we made it. When we stopped at the first red light the mickey exploded. Wow! We figured the store employees drank most of the vodka and topped up the bottle with lemon-lime soda. At least it didn't stink up the car, but the guy decided he would avoid vodka grenades in the future.

It may sound like we were "hard at it" but these are just anecdotal incidents. We were experimenting – seeing how far we could push it without causing a major ruckus.

We all had bogus Ontario liquor permits – little yellow wallet-sized deals with name and age written in ink. They were easy to get, but using one was stressful. If the clerk looked you in the eye, certain muscles tightened up and a week's worth of constipation was in store.

One Saturday night three of us decided to be men and have a cool one at the Old Fort Frances Hotel. We'd picked the night and the location intentionally, knowing the beer parlor would be busy and also knowing that the waiter on duty didn't know us. We sat in the darkest corner, kept our mouths shut and our eyes open. Our main concern was that someone in the bar might know our dads and if ratted out we'd be in big trouble. It turned out to be no fun at all. We left after one round.

Sidebar: How attitudes have changed! More than thirty years later the wife and I added a liquor and beer sub-agency to our C-store/gas bar. Every Friday and Saturday night we continually carded under-aged kids and drunks, and they had no qualms about digging in their heels. It was a continuous hassle and I never liked that part of the enterprise. (But – as our District Manager told us – selling booze is the only business where you can piss a guy off today and he'll come back tomorrow.)

Table Scrap: Smart Cop Story. (This is out of context, but it <u>is</u> liquor sales related)

One weird aspect of human nature we soon learned, was that a seemingly sober person, once they had made their booze purchase would have an immediate attack of rubber leg syndrome, as happened one evening.

He looked to be totally unimpaired until he picked up his 24-can case and then he started visiting and babbling with other customers in the store. I waited for a chance to quietly usher him out, but we were busy. A lady came to the pumps and she was known as a no-nonsense type, so I hustled out to pump her gas. Maybe she would take 20 bucks worth and leave – no such luck – she went into the store.

Now a cruiser pulled up to the pumps. We appreciated their business and visibility, but this was getting dicey. We always guarded our reputation and I just wanted a chance to head the drunk off myself. I knew I could find someone to drive him home, but the lady came out and told the cop that the guy was drunk as a skunk and should not be driving.

The cop handled it well. He knew the keys were in the pickup, but he stopped the jerk before he reached the driver's door. I jumped into the fray and said ---- ---- would drive the guy to a party house one-half-mile north which ---- ---- gladly did.

The cop followed them, and at the house he threw the truck keys into the bush.

Good move – the roads stayed safe that night and I don't think they ever did find the keys.

By Christmas I was sick and tired of banking – I had lasted eleven months. My brother Rod, who was now a pilot in the Royal Canadian Air Force, (RCAF) was being posted to Trenton, Ontario, and he made me a proposal. He was determined to make an over-achiever out of my worthless hide, and would bankroll me to attend high school in Trenton. I could pay him back when I became rich and famous. (I never did become R&F, but I did pay him back a couple of years later.)

But there was no room in the budget for the white Monarch – I had to sell my baby. It went to a guy further down the Rainy River Valley.

Fifty-five years later I was retired, living in that community and an old friend told me that the Monarch was still sitting in the bush about four miles east. He said it looked pretty sad – rusting away with a huge poplar laying across the caved-in roof. I borrowed my step-daughter's digital camera and drove over to where the car sat a hundred yards behind a hunt shack. I parked in the driveway and sat there for ten minutes or so – did I really want to embarrass the old girl? I decided I could not do so – let her and my fond memories remain as happy and shiny as those days so many years ago.

I went home and gave the unused camera back.

Chapter VII   Fiddlefoot Training – Senior Year - 1959

In Trenton I stayed with a nice little old lady about five blocks from the high school. The school was a fairly new, modern facility and as the town and the student body were a mixture of kids from multi-generation local families and constantly revolving air force personnel, I had no trouble fitting in. I also had Ron and Patsy to visit – two cousins who I had not seen since the early '40s in New Brunswick. Back then Patsy had been a freckle-faced redheaded tomboy – a complete pain in the ass. Now she was still two years older than me but had magically turned into a self-confident beauty. I should never have told anyone she was my cousin – I should have pretended she was arm candy.

Unfortunately, the new me was still old me. The school/social mix was still 25/75 and I finally convinced my brother that I was not cut out to be an academic.

Sidebar: looking back though sometimes faded panes of memory windows, those six months at Trenton still come through strong and clear. Because the town, my fellow students and Rod were kind and non-judgemental, it was like an oasis, or a still pool at the foot of rushing rapids. For six long months I could stop paddling against the current of a turbulent world, and, like a Buddhist retreat, it prepared me for fast water in the future. I now think that Rod and Trenton may have saved my life.

Near the end of June I packed all my worldly possessions into my dad's old Gladstone suitcase and hit the two-lane blacktop with my thumb out. I had no firm plan, just a vague idea that perhaps I would go west to find work. Hitchhiking was no problem in 1959 – I had little competition out there. The roads would not be crowded with flower children until the '60s.

I reached the farm in two days, checked in, hugged my mom, showered and two days later left for Winnipeg, now carrying a duffel bag along with my suitcase. I bunked in with my sister and brother-in-law and looked for a job. I found one the next day, by gum – selling Goodyear tires in the heart of Winnipeg. I would start next Tuesday.

I also connected with a couple of old high-school chums who had a loft apartment near downtown. I moved in with them and went to sell some rubber Tuesday morning.

After a quick tutorial with another new employee, the sales manager turned us loose to sell, sell, sell. We were told there was a broad price range – list price, (if we could get it) ranging down to the bottom line if the customer was tough. Mounting and balancing could be used as bargaining tools and when we neared the bottom line we were to pretend we had to have the manager's approval. I never was too good at telling lies. I could sell tires, though, but there was a fly in the ointment. Three days later I would discover, much to my surprise, that I did have some principles after all.

I sold a tire or two at various profit margins to various savvy and semi-savvy customers. They all knew that the list price was hogwash and while most were merely being thrifty, some were downright ignorant, knowing full well that by being nasty they could drive this hick to his knees. Another thing was that if I and the other new guy were waiting for a prospective customer to come in it was a foot race. I preferred to amble but he was a dasher, and one day he straight-armed me on his way past to get to the guy first. It turned out to be the mailman and I got a good chuckle out of that one.

Friday afternoon I met my Waterloo in the form of a nice young farm couple. They wanted four new tires for their ¾ ton farm pickup and they called me "Sir!" The truck sat in the lot – it was clean and neat, but far from new. The young folks were also clean and neat, and they were very new, if you can catch my drift. In a matter of minutes, I had them into top-of-the line tires at list, and in my enthusiasm I even tacked on mounting and balancing.

This was far too easy. These young folks were so nice, so respectful, and they were treating me a whole lot better than I was treating them. I took a break from my sales prattle to consider the situation. These were good people – not flush with ill-gotten gains – just starting out in married life and no doubt soon to have a mortgage and children.

I got a fit of the dreaded ethics. I moved them away a bit from the sales manager's office and in a lower voice I told them I would cut the price by five bucks a tire and throw in balancing and mounting. They beamed and signed the contract. I took the contract to Mr. Manager for his initials and the sharp-eared bugger had overheard me. "We will talk later," he said. (Dum-da-dum-dum!)

That Friday night we had a small party – just the three of us with a twelve-pack of beer and a bottle of wine. We played some cards and talked about the good old days – three guys who were not yet old enough to legally buy the alcohol we were drinking, talking up the good "ole" days! The other two didn't have to work Saturday. I did, and I slept in!

I didn't even bat an eye, nor did I call in with apologies. I didn't like the manager, I didn't like being beaten down by jerks and I really didn't like racing Myron to the next mailman. I liked tires though – tires made my world go 'round.

The me vs me option consideration didn't take long. I already had 1400 miles on the clock. The coast was clear and the West Coast beckoned – it was Vancouver or bust. My roommate Tom came aboard – he didn't like his bank job much, either. The only question was; would we hitch or drive? We decided to drive.

Tom had a '52 Meteor in a body shop north of Winnipeg. He had flipped it in a soft snowbank last winter with not much more than a few wrinkles which had already been dealt with. It was waiting for another cash infusion to do the paint job. Neither one of us had any money so it was decided to retrieve the Meteor, trade it down for a lesser car and a little cash for gas money.

We took the Meteor to an auto wrecker on the north edge of Winnipeg. He was pretty sure we weren't crooks, but he was also pretty sure we were feeble-minded. We drove away at 4 pm in a lesser car and a little cash for sure – a 1947 Chev four-door and a hundred bucks. We bolted the Meteor plates onto the Chev and we split the money 50/50 – I would repay Tom later.

The Chev was a work of minimalist art. Originally dark blue, it had been painted a bright green below the beltline – with a brush! The best that could be said about it was that it had all it's windows and four tires – only four – there was no spare. Only three tires had any tread – the other was as bald as a drag strip slick. We soon found out why, because the first time we hit a red light that wheel locked up.

We packed up and left Winnipeg at 6pm. I took only my duffel bag containing clean underwear, tee shirts, socks, an extra pair of jeans, my electric shaver, tooth bush and a blanket. (I would be glad later on that I had that blanket.) For possible cool weather I also had my brother's university jacket.

I called my sister. I knew I'd get a blast, so I kept the conversation short and one-sided. "I'm heading for Vancouver, will you pick up my suitcase when you get the chance? Bye." Click. Did Tom call his folks in Fort Frances? If so, it must have also been a quick call.

We headed west on Portage Ave, but after a couple of blocks we pulled to the curb for a short chat. Straight west would lead to Vancouver and a left turn would (could) lead to Mexico. We flipped a coin and Mexico won!

We reached the border at Emerson with a few daylight hours to spare. The customs officer looked at us, at the car, and raised an eyebrow. I took it as a good sign – two eyebrows might have meant trouble. We told him we were going to visit a couple of girls in Grand Forks, N.D. and we would be returning later that evening. He told us to drive carefully. "We sure will, sir."

We found that 50 mph was the optimum speed for the Chev. Any slower or faster and a loose wrist-pin got real noisy – at 50 it behaved better. At nightfall we discovered our headlights were rather dim. A belated safety inspection told the tale. We had only one headlight on high beams and on low beams we had no lights at all!! We decided not to test the highway patrol so we pulled off at the next decent spot and slept in the car.

Other slight problems showed up over the next 24 hours, one being that the vacuum shift assist had packed it in. (They all did by 1947 ½.) If you failed to shift it out of gear before a stop sign it took a bit of a push by a passenger to get it into low gear. We ran stop-signs on upgrades.

We hit the road again at daybreak, happy and gung-ho. We talked about picking up hitchhikers – we hadn't passed any yet but we decided that the odd different point of view might be interesting. Somewhere in South Dakota it looked like rain and we picked up the first guy we saw. This would later prove to be a big mistake.

When it started to rain another Chevy idiosyncrasy became evident. The windshield wipers had rubber only here and there, and the vacuum was rather feeble. It started to pour and the highway ahead and oncoming traffic looked like they had been painted by Dali. I pulled over until the rain let up and it was the last rainstorm we would encounter.

(I was doing all the driving, as both Tom and I knew that his driving skills were lacking and I was not about to let the hiker drive. He was already exhibiting his own skills – mainly bullshitting.)

We pushed it a little too late that day. In Kansas darkness caught us before we could find a suitable parking spot and a state trooper pulled us over. He was a nice, older gent, and asked us where we were from, where we were going, and did we know we only had one headlight? We told him we were from Canada and we were heading for the Gulf of Mexico – we had never swam in salt water. As for the headlight, "Gee sir, we're glad you pointed that out." (We had already discovered that we had one light only on high beam – and <u>no</u> lights on low beam.)

"Try the dimmer switch," he said, and walked to the front of the car. I shut the lights off and quickly turned them on again. "Yup," he said. "Only one on brights, too." Phew!! We were lucky it was dark and he had no flashlight in his hand. Had he gotten a better view of the headlights he would surely have noticed that there were no bezels around the lights and each sealed beam was held in place by <u>one</u> little screw at the top. Every bump on the road made the one working headlight wink at oncoming traffic.

Then he checked the tail lights – something we had failed to do. Wow – we had two taillights, but only one brake light. "Better get those lights fixed tomorrow, boys." "Yes sir – we sure will."

Now he looked at the tires and told us one had no tread. I told him our spare tire was flat and would be fixed tomorrow, and bless his heart, he didn't ask to look in our spareless trunk. He left us with a suggestion that we should park for the night as soon as possible, which we did.

On the road at sunup again and around noon, just before Omaha, a front wheel bearing started to get angry. Another junkyard came to the rescue – fifty cents for a used wheel bearing. He lent us a jack, a crescent wrench and a pair of pliers, and thirty minutes later we pulled out of Omaha with the broad Missouri and Council Bluffs, Iowa on our left.

We were making good time. Not fast time by any means, but we never gave Chevy a rest. She was easy on gas and oil and when we gassed up we grabbed a chocolate bar and a pop, our main diet to date. The car had to be fed but we could live on stored fat.

The hiker was already starting to bug me. Tom and I were enjoying the scenery. There was stuff new to us passing by – big and little rivers crossed on big and little bridges. The agriculture was changing to unfamiliar things. Was that a cotton field – that looks like peanuts, but every time we marvelled at things, the hiker had been there, done that, and had seen something better. Soon Tom and I shut up – the BS'er in the back was taking the fun out of the trip. What was more irritating, was that he had commandeered the back seat and the night before Tom and I had slept sitting up in the front. I was hoping we would soon drop him off at his destination, but it seemed he was going where we were going. Why would he split? He had a chauffeur and a captive audience who bought him chocolate bars. He said he was broke, and maybe he was, but we were running low on cash. Every bar he ate and every pop he drank was cutting into our measly bankroll.

Today we planned to hit the Gulf before dark, but we passed a sign – KANSAS CITY NEXT LEFT – and we turned left.

We could not resist it. We were going to Kansas City – Kansas City here we come! We had heard that there were pretty girls there and we were going to get us some. Well – the first pretty girl we saw in Kansas City was at the first stop sign we hit. She was in the right hand lane in a '58 Impala hardtop. Cool dudes us – we snubbed up the Chevy with our bald tire sliding and she gave us a quick once-over and a sneer. We pulled a U-turn and left town. We'd been there and saw a pretty girl but we had a date with a sand beach at Galveston.

That side trip messed up our schedule. We had planned to hit the beach before sundown, but as we passed Nacogdoches it was coming dark and before we reached Houston it was pitch black. Once again we were running on candlelight and although we had been trying to avoid busy roads we were now on a four-laner and it was packed! I thought if I stayed tucked in between two highway haulers in the right-hand lane we could escape scrutiny. No such luck – a state trooper nabbed us again!

This was a young fellow, very spiffily turned out with that oh-so-perfectly-brimmed Stetson which state troopers still wear and which still impresses me no end. He had a pure southern drawl – "Y'all goin' t'Hooston?"

I didn't understand a word. "Pardon me, Sir?" He obviously had the village idiot to deal with. "Are—you—all—going—to—Houston?" – perfectly enunciated.

"Oh, Yessir! For sure, sir! In fact, my uncle lives just off the freeway just ahead, sir." I babbled.

We heard his radio crackle in the patrol car behind us. He had to go. "Be careful," he said, and trotted back to the cruiser. We had dodged a bullet again.

Sidebar: There is no way, no how, that anyone could do today what we did in 1959. We were driving a hunk of weirdly painted junk with poor rubber, poor lights and we were in Texas, and we were almost cashless. We had false Manitoba plates on a car driven by a guy with an Ontario licence and not once had we been asked for ID or our non-existent Chevy registration. Not once had we or the car been searched and our spareless trunk was never looked into. We were so far under the radar that gophers were ducking as we cruised by. This would be the last time that a cop stopped us. How in blazes did we avoid the slammer?

Sidebar: We only had Canadian funds, but it never was much of a problem. Our money was pretty close to par then, and we cashed most of our twenties and tens before we got too far south. Even when we reached Texas few problems arose. Sometimes a clerk might study on the bill and the trade it for US paper from his own pocket. One clerk wouldn't take a perfectly good dollar. A guy behind me did, giving me <u>two</u> bucks.

On toward Galveston we drove and passed a service station with gas at 21 cents a gallon. Dash it all, we had paid 25 cents fifty miles back! Had we run Chevy a bit lower we could have saved six chocolate bars.

Galveston was positively eerie. Refineries to our left and right spewed smoke and flare stacks flickered through the dense atmosphere. Through our open windows wafted the smell of Texas Tea.

The four-lane morphed into a little-used two-lane which ended at the beach. It was deserted and an onshore breeze kept it smog-free. Our one-wheel brake dug in and before the car came to a full stop Tom and I were running to the surf, shedding clothes along the way. We plunged into the water with a "Whoop!" We had hit the Gulf of Mexico! (Meanwhile, back at the ranch, Mr. Hiker was sitting, unimpressed – been there, done that.)

As we splashed and swam in that warm water I told Tom it was time to cut the leech loose. The trouble was, although we were both polite Canadians, Tom was far more polite than I, and he thought it was harsh to abandon the guy. Since I was travelling in Tom's car on money borrowed from Tom, I acquiesced.

I was a bit disappointed when I got out of the water. I expected to feel fresh and clean like when I had a dip in the lake at home. No one had ever told us that one should shower after a salt-water swim. (Mr. back-seat-know-it-all took great pleasure in telling us that he knew it all along.) At least we had sluiced off some of the dust from the Chevy which had been collecting same for years in the Winnipeg salvage yard.

Now I was going to have a good night's sleep under the stars. I spread my blanket on the beach and was once again disappointed. Within an hour sand fleas were sharing my blanket and my blood – it was back into the front seat for me.

I woke up in a poor mood and we counted our pitiful cash reserves. There was maybe enough left for two more tanks of gas and we had to have a hamburger for brunch. We did so on our way to Corpus Christi – our first actual food since leaving Winnipeg.

Why Corpus Christi? Because, for once, the hiker had some constructive input. We needed to find work and not just any old job would do. We thought it might be interesting to go to sea, and the hiker said that CC was a merchant marine centre.

He was right – we hit the waterfront in Corpus Christi and a big sign on a warehousy-looking building said something like "MERCHANT MARINE CENTER, SAILORS-R-US." We went in, found an office and we also found out that they didn't give a rat's behind whether we were legal or illegal or whether we came from Canada or Katmandu, the high seas were an Equal Opportunity Employer. (We also found out that the hiker claimed to have no ID – o and we thought WE were under the radar.)

Sure we could sign on to a ship. The only catch is that we had to take a ten-day seafaring course and even that was no big deal, although slightly convoluted. The course was free, but to take the course we first had to find a ship's captain who would take us on. Even that was fairly simple. Captains knew the drill and though they might be pulling anchor the next day, all they had to do was give us a piece of paper signifying that they were wiling to hire us. Then – after completing the course we could nab any other ship in port that needed strong backs to pull the oars. Weird, but that's how it worked.

Tom and I talked it over. The stickler was how to survive for two weeks while taking the course. We didn't have enough moola to eat for that long and if we sold the car for eating money we would have no place to sleep, and also, we were running out of clean clothes. We decided to try for El Paso. Maybe we could get to see Mexico before we starved to death.

We shut down for a night's sleep at Sonora and left the next morning on our last tank of gas.

Heading west, it got hot – hotter – hottest, and the wrist-pin really started to complain. I stopped and experimented, pulling off plug wires until the hammering stopped. We tried the Chev on five cylinders, and although it ran a little rough, the wrist-pin was silent. The temperature gauge rose and I kept an eye on it.

In the back of my mind I was studying on how to survive. We were in mainly farm country with some fruit and nut (pecan?) groves. There were things a farm boy could do. Tom was a town guy but I knew I could cover for him, and given a chance he would learn how to dig in. The unknown cipher was the hiker – he had already proven to be as lazy as a cut cat and I was unwilling to embarrass myself by foisting him on some hard-working farmer.

So we rode along with Tom and I chatting about alternatives. Tom had an uncle in Portland Oregon. If we made it that far, maybe Unc would bankroll us to Canada. For a change the hiker in the rear was silent.

Southwest of San Angelo the landscape changed dramatically and it was as beautiful as it was unexpected. I know now that we were coming off the Edwards Plateau, but all I knew then was that spread out below us was a vista of semi-desert as

far as the eye could see. Cactus, sagebrush and clumps of grassy vegetation were scattered about along with a cow here and there. Beside the arrow-straight road off to the west, we could see a water tank and a windmill. Another one was tinier in the distance, and these water tanks would save the Chevy's life.

We stopped, got out and surveyed the view with pure, unadulterated enjoyment and then the hiker ruined it all – he'd seen better. I was so pissed that I could have thrown him over the edge.

We drove down the switchbacks to the flatland and now it was really hot – which surprised me because I thought it couldn't get any hotter this side of you-know-where.

The Chevy started to complain and what looked like heat waves coming off the hood was actually steam from the rad. The heat gauge had red-lined.

We made it to the first water tank and after a few trips with a tin can left beside the tank, the rad got a transfusion. We kept the can and drove on, much slower now – and the temperature gauge was starting to climb again.

The water tanks were spaced twenty miles apart and I guess the Chev and the longhorns had the same grazing perimeter. All afternoon it was drive 20 miles, top up and repeat.

Past Fort Stockton it was coming late afternoon and the day was cooling off. A hundred and twenty miles from El Paso the hiker yapped some stupidity crap and I'd had it! At a junction with a highway leading north, I pulled over, got out, and yanked my duffel bag out of the rear door. Tom was aghast – He wanted to know what I was up to.

"I'm splitting," I said. Just that – "I'm splitting."

Tom got out and we had a conference a few feet away from the car. He, of course, wanted me to stay, but I was adamant! I knew if I stayed with them someone would get the crap beat out of them, maybe me. Tom asked me how much money I had left – I told him $1.10. Tom had a bit more than five bucks. He tried to give me some but I told him he had that pecker-head to feed and if he dumped the guy and cut his losses he could sell the Chev and make it to Portland. Tom had tears in his eyes as we shook hands goodbye, and I almost – almost caved in.

(The next time I saw Tom was in the early '90s when Fort Frances hosted a huge high school reunion. He told me he'd sold the Chev in El Paso for ten bucks and had bought out the hiker's option for five. He hitched up through California to his uncle's, phoned home and his parents sent him a plane ticket. He later on became a successful accountant (?) in Calgary and – in El Paso he walked across the Rio Grande, turned around and walked back – he had made it to Mexico.)

I stood on the side of that two-lane blacktop and watched the Chev fade off into the distance. I looked east back toward Fort Stockton and North towards New Mexico – no traffic in sight. My watch told me it was 6:30 pm and my tummy told me it was suppertime. The day had cooled to a pleasant July evening. The air was still and there was an occasional bird call. Here and there among clumps of low bushes stood the odd tall cactus with semaphore arms. The road was lonely, the countryside was lonely, and suddenly I was so lonely! I sat on my duffel bag and bawled like a baby.

I was not feeling sorry for myself (well, maybe a little bit.) Mainly I was sorry I'd let Tom down. I was sorry I had not insisted on kicking the bum out of the car. I was sorry that my habit of shooting from the hip had put me (and Tom) into this predicament. I was a little sorry that I only had a buck ten in my pocket, but for some reason that didn't really bother me. Mostly I was just sorry.

It was a short and cathartic melt-down. I stood up, hitched up my jeans and got on with it – mainly standing there wondering if the end of the world had stalled all the vehicles.

I decided to thumb the first unit no matter which direction it was heading. I hoped it wouldn't be east – There was nothing that interesting about Fort Stockton. West? Maybe – but what if I passed the slow-poke Chev and thumbed them? They might give me a one-finger salute and drive on by.

A truck appeared from the east and the signal indicated it was turning north. I hustled across the road, he stopped and I was riding my thumb to New Mexico. (Thirty-four years later I told Tom that while he had made it to Mexico I had made it to <u>New</u> Mexico.)

The truck was a green pre-1954 GMC stake body with a load of new tires. It was immaculate, looking like it had just been driven out of the two-ton showroom. The driver was a black guy in his early thirties – also immaculate in his work duds and very friendly.

The conversation was the usual stuff. "Where are y'all going?" "Why are y'all going there?" "Where did y'all come from?"

It was a two-way questionnaire to establish boundaries and to assess each other's character, but while he was coming across as a hard-working family man, I was more likely being judged as a foot-loose flibberty-gibbet, obviously a little off-centre.

He asked me where I would spend the night. I said I would just spread my blanket on the desert sand – he was shocked. "Those rattlers will eat you ay-live!" Brrrr – point well taken.

I promised him and myself that blanket-spreading would wait until I got out of snake country. (Later on I would land a job and for the rest of my trip I would generally have a hotel room for sack time, but one night after staying out too late waiting for a ride that never came, I <u>did</u> crawl into a dusty truck seat in a quiet compound, figuring snakes couldn't open doors.)

Table Scrap

I already had a bit of hitching experience. I was sixteen, I think, and in late August, mad at my dad, (or vice versa) I had thumbed to Saskatchewan, shovelled grain for two weeks and returned home in time for school. On that trip I spread my blanket a couple of times – once in a quiet pasture and I awoke at daylight enveloped in ground fog. It was so quiet and smelled so good – and as I lay there enjoying nature there was a thunder of hooves – STAMPEDE!

I hopped up, looking for high ground, but it was just a herd of horses – curious horses. They stood in a circle, wondering who let this interloper into their private park?

And on that hitchhiking foray I learned one constant truth – 99.9% of people who pick up hikers are darn nice folks. Heading home through Sask, I had been picked up by an old farmer in a '53 Chev. He was very friendly and very talkative, and we traded stories as we drove along – his much more interesting than mine.

It was a short drive but a long one. The old fellow was so happy to have an attentive captive audience that he forgot to shift the car into high gear. I didn't want to embarrass such a nice gent by pointing it out, and when he slowed to drop me off he was quite surprised to find his shifter was still in second gear. We both chuckled.

The black guy was going to Carlsbad. Before dark we crossed a bridge over a fairly wide non-river. I asked him why anyone would build a bridge over a dry river containing nothing but sand.

He said that sometimes there would be a downpour in the low foothills to the northwest, and even though it might be sunny here, the river would be full of muddy, rushing water. Pretty cool, I thought.

At Carlsbad I waited quite a while for my next lift. I had heard about the caverns but my budget was not up to touristy stuff. Finally, well after dark, I got a ride to Roswell.

I was dropped off after midnight at an old-timey 24-hour Texaco station. I washed my face and hands in the washroom and asked the night attendant if I could spread my blanket in the service bay. He wasn't fussy about it, probably thinking I was an alien. Well – I was an alien, but not an Area 51 type.

I was whipped, and ignoring his protestations I bunked in beside the grease pit and slept like a log, arising refreshed but hungry at 4:30 am. Dawn would soon be breaking and I had many miles to go.

To reward the guy for his semi-niceness I bought breakfast – two Snickers and a Coke. I scarfed a free road map and hit the highway. The attendant actually smiled and wished me good luck.

150

Sidebar: I think it is time now for a short disclaimer. It's more than 55 years later as I write this, yet I have almost total recall of that summer so long ago. I kid you not. I <u>was</u> there, and I <u>did</u> that.

Rides were few and far between out of Roswell. Vacationing families don't usually pick up hikers. If a car or station wagon contained more than two people, I kept my thumb at my side. Farm folks were my best bet. I hit Tucumcari just after lunch time, which I didn't have because I was harbouring my last 70 cents. I was dropped off in the middle of town and I walked east. It's a good thing towns and cities were smaller in 1959. I would often have to walk to the edge to hitchhike effectively.

My map showed me the way. I could cut up through the corner of Texas, through the Oklahoma Panhandle and into Kansas. Farm country and a possible job lay ahead.

I think it was on my second ride out of Tucumcari that I hit paydirt. The car was a

1951 Ford convertible, top down, black with brown leather and another immaculate showpiece. The driver was near my age and on his right sat a young serviceman, and I got into the back beside the soldier's gear. This driver was obviously a hitchhiker picker-upper.

Pure Nirvana – I leaned back with the warm wind whipping my hair and with a panoramic view of all things roadside without an interrupting doorpost. The two guys in the front chatted away, turning once in a while to bring me into the conversation. I couldn't hear a word – I just smiled and smiled. On the floor beside me sat a briefcase. The ID card said, "John Enns." Very interesting – was John Enns the driver or the passenger?

Sidebar: In the Manitoba Interlake during my elementary years our small underfunded school board could only afford "supply teachers," who either planned to attend teachers' college or were halfway through the two-year course. We had a different teacher every year, one of whom was John Enns.

The soldier got off at an intersection and I moved to the front seat. The briefcase was still in the back.

Now we could hear each other and we shared story lines. He was attending theological college in California and returning home for the summer break. I was a Canadian recently out of school myself, and investigating cross-border culture. I took the opportunity to say, "I notice your name is John Enns. I had a teacher in grade 6 by the same name – could you be related?" – and just like that, I was in like Flynn.

(I knew darn well that there was little chance of a relationship between the two Johns. In the Mennonite community the surname Enns is as common as Smith in Ontario or Tremblay in Quebec.)

The young fellow was delighted – it was like old home week and home was in Liberal, Kansas. Bonus – Kansas was wheat country.

At Liberal I was introduced to Mom, Dad, and various members of the Enns clan. Mom took one look at me and it was tubby-time. While I bathed and shaved she washed and dried all my clothes – all of them – even my old army blanket got the treatment – only my shoes and work boots escaped the machines.

As I dried myself with an oh-so-soft-and-fluffy towel I caught my reflection in the mirror. I had been driving for five long days with my left arm on the Chevy windowsill. That arm was a nut-brown and my right arm was drawing-room white. No wonder the Roswell gas jockey thought I was extra-terrestrial.

Then it was suppertime – and what a supper! I tried not to wolf my food, but with so many people talking about so many things I didn't have to join in the conversation. Platters, plates and gravy boats were continually making the rounds – second and third helpings were expected. Then came apple pie which I passed on with regrets. I was fit to bust!!

The house was crowded but somehow a bed was found for me. Full as a tick, with my squeaky-clean carcass between clean sheets on a soft bed, it was a peaceful, dreamless night. (Mom never mentioned that I had 70 cents left but she must have known – she had emptied my pockets when I was in the tub.)

At breakfast Mom was very, very apologetic. What I had thought was an impromptu family gathering was in fact the leading edge of a reunion. More people were coming and there just wasn't room for me. I told her I understood completely and not to worry, she had done enough for me. I gave Mom a hug and shook hands with Dad. Dad gave me the phone number of a custom harvester in Kimball, Nebraska and John drove me back to the highway.

That son-of-a-gun offered me money – so Mom had ratted me out after all! I respectfully declined the offer, telling him my folks would wire me cash if I didn't get a job. (It was an white lie, of course.) John wished me luck and told me to keep in touch. To my everlasting regret I failed to do so.

I hitched north towards Ogallala, totally clean from head to toe. I had been well-fed, I had a phone number in my pocket and I had a renewed confidence in the kindness of strangers. I was ready for anything – onward, onward, onward.

Table Scrap: Hitchhikers' Handbook

Rule 1: Always appear as neat and clean as possible. No suits or sports coats – they look downright fishy. I only hitched in the summer so jeans and a white tee shirt did the trick.

Rule 2: If you have a prop, use it. My brother's university jacket served me well. I carried it over my arm with "Western" clearly visible and since I was in the western US, "Western" was innocently generic. One guy did ask me where Western University was and I told him it was in Eastern Ontario. How oxymoronic!

Rule 3: Never tell an outright lie if you can avoid doing so. If the ride is long the lie can catch up with you making both you and your benefactor feel bad. White lies are OK – they save face all around. For example: "Just out of school, hey?" "Yup," - truth. "Going back this fall?" "Yup – if I get home in time." Outright lie – I was done with school. The "if" modifies it to a white lie.

Rule 4: Never thumb a vacationing family, just smile and wave. With two or three kids and a dog in the back seat they already have enough problems. The only reason dad might stop before dark is if some kid or animal is in danger of peeing themselves.

Q: How did Lewis and Clark make it to Oregon on schedule?
A: Dad was driving. (Credit – Dave Barry, Miami Herald.)

Rule 5: Never bum money, no matter how broke you may be. If some guy offers you a buck or two, refuse politely. If they stuff it in your short pocket, accept it. Your pride is intact, and you don't want to hurt the guy's feelings.

Rule 6: Be helpful if at all possible. I have been a substitute driver when asked or if circumstances indicated. On my harvest hitch when I was sixteen, I helped hand bomb a trailer load of cement bags. The consignee gave me ten bucks which I took because I had earned it – cement bags are heavy!

Rule 7: Never, ever give a non-stopping vehicle the middle finger salute. Thumping may occur.

Rule 8: Never brace a trucker at a truck stop. You are putting the guy on the spot, and if he does let you in it may be a surly trip. Stand at the truck stop exit. The guy has to slow down to enter the highway anyhow.

Rule 9: Never toady up to a driver. Harsher critics might say that I was sucking up to John Enns, but I was just opening a door. Hikers don't toady – hikers are opportunists.

Rule 10: Never disrespect the driver and/or his/her wheels. This actually happened years later when I picked up a hiker. He was anti-everything. I shouldn't be driving a gas hog and I should be taking better care of the environment, and on – and on – and he was still carping when I pulled over and hit the eject button.

Rule 11: Never offer the driver gas money. Offering to pay for a ride changes the dynamics, making him/her a taxi driver. If asked for gas money, beg penuriousness, and even if you are flush, don't let on. Sometimes an exception to rule 3 is called for.

Rule 12: If a family picks you up, be nice, pet the dog, chat with the wife and kids. Don't light up – even if the driver does. Never roll down the right hand window to stick your arm out without asking permission. Don't fall asleep. The driver may have been tired and picked you up to help keep him/her awake – common sense stuff.

Rule 13: If a weirdo picks you up – bolt! Don't go out the door at 60 mph but bolt at the first chance. Just bolt and don't look back.

So I made my way to Ogallala and turned west towards Kimball. There were no stand-out rides, mainly the shorter variety with the usual nice folks – no chocolate bars or pop, either – I had grubbed up enough groceries in Liberal to last me for a week. I did keep hydrated, however. Coca-Cola, "The Pause that Refreshes," had a machine at every service station with a cold water tap and a Dixie Cup holder. What a life-saver on a hot July day. Darkness caught up with me somewhere east of Kimball. (Memory jog – this was where I slept on the dusty truck seat.)

The next morning, I reached Kimball around 9 am. This was where I was to contact the custom harvester. I was dropped at an intersection and about a mile west I could see the town. To the north was a farm-yard just off the highway. I figured the custom guy would not live in town, so I decided to see if I could use the farmer's phone.

The wife answered my knock. Slightly surprised, she wondered if I was selling something, so I explained the deal – I wanted to make a call to the custom harvester. She knew the guy and just as I expected, he had his own farm. It was a local call and she was glad to let me use the phone.

The harvester was also nice. Yes, he needed men, but the combines were not rolling yet. I was a bit disappointed – I had been passing golden wheat fields, but I guess they were not yet pure gold. He said that if I could hang around the harvest would kick off next week. I said I was heading north, hoping to pick up a little work on the way, and he gave me his itinerary adding that if we reconnected, he would take me on as people tended to drop out in the earlier part of his northward trek. I walked back to the highway confident that I could find something to survive on until the harvest started. This part of the western U.S. was brimming with good people.

The first vehicle heading north was a newer single-axle GMC grain truck with an old gent driving, and following him was an older but nice looking Reo Gold Comet. The young fellow behind the wheel pulled over and my duffel bag and I jumped in. (Throughout that whole summer, other than our junk-yard Chevrolet, I never saw or rode in a shoddy vehicle – never.)

We exchanged info. He and his two brothers farmed with his dad who was in the GMC ahead. They had just delivered two loads of barley to Kimball and were returning home to their farm southwest of North Platte. I told him I was a Canadian seeing new country and looking for work along the way.

He introduced himself – Tom Van Pelt – and I unabashedly played the name game. (See rule #9) I told him that a couple of years ago a Jim Van Pelt had quarterbacked the Winnipeg Blue Bombers, a team in the Canadian Football league (CFL.) Was there a connection? Of course there wasn't, but it was an ice breaker.

He became much more friendly now. I was looking for work? His older brother had recently been thrown from a horse and had broken his arm, leaving them short-handed. He and his dad were planning a coffee break at Gerling, just south of North Platte. He would talk to Dad.

He bought me a coffee and told Dad that I was a farm lad and available to help out. His father, a quiet, calm old gent, agreed to give it a shot, and just like that I had a job! We headed west on a broad gravelled road – I was pumped.

(Can you believe it? I had made 900 miles on 40 cents! Chew on that, you eco-friendly Prius people.)

It was a 20-mile trip to the farm and I got an operational overview on the way. They had 40 acres of irrigated alfalfa almost ready for a second cut. They had eighty-some head of purebred Herefords and they raised and saddle-broke quarter horses. No – I would not be expected to ride a bucking horse – I might break more than an arm. They also grew oats and barley for stock feed and a good year would leave a little left over to take to Kimball. It seemed to me to be a pretty small deal to support his parents and three families. I would soon learn I had woefully underestimated these folks.

This was rolling hills country. We pulled into the home place where Dad's house stood on a low hill surrounded by garages and a couple of grain bins. Over to the west was a pasture and corrals. Not far north of that was a horse barn, another corral and a horse pasture. Nearer at hand was a machine shed and a barn. Some implements sat outside. One was a hay wagon with something attached to the side that looked like a small Ferris Wheel. Downslope north of the machine shed I could see what looked like an old threshing machine and an old square-cabbed grain truck – derelicts, I mis-presumed.

Dad's house was a new, but not fancy, midsized two-bedroom bungalow. Not far off but far enough to give some privacy was his eldest son's similar house with an attached garage.

What wasn't new was well-kept and neatly painted. In the yards were ornamental shrubs and the odd tree. I was used to seeing shelter belts on the Canadian prairies and I rightly assumed that in Nebraska's shorter winters they could put up with some cold winds, preferring bug-clearing breezes in the long non-snow seasons. Tom told me that he and his brothers were the third generation on the place. It had been homesteaded by Grandpa and Grandma in the previous century.

So it didn't look like a big hairy deal, but one thing intrigued me. As we put away the grain trucks, each in its own garage, I noticed three 1958 GMC 3/4 ton 4x4's. Hmm – maybe this stillwater-farmstead ran deeper than I thought.

I met George, the youngest brother and a nice guy. The older brother was not around – he was at the doctor having a new cast put on his busted arm and I would meet him tomorrow. He would prove to be the family money-handler and not a nice guy at all.

We went to the house where Mom had a mid-afternoon tea break laid out – cookies and cake. It sure tasted good.

I learned I would be getting eight dollars a day with bed and board – not too shabby. Tom took me home in one of the pickups to drop off my duffel bag. He lived a few miles away on his own quarter. (George also had a house on his own land.) On the way to Tom's we passed fields of ripening wheat and here and there rocking horses were bringing oil to the surface. Someone was doing ok, I thought.

Home was another newer bungalow. I met Tom's wife Devoda and two young pre-school children, and I was shown my room in the finished basement. It looked pretty comfy and I even had my own bathroom and shower.

But the plot was thickening. We had parked beside a brand, spanking new white-on-white 1959 Oldsmobile Dynamic 88 four-door hardtop. As I walked past the car I glanced in and spotted power window buttons and factory air vents – this baby was loaded. I thought maybe it belonged to a visitor but no one was visiting – hmm, and hmm again.

I changed to work boots and it was back to the ranch to move irrigation lines, which were on wheels but not self-propelled. In the centre of the field a big Waukesha V-8 pulled water from the Ogallala aquifer and fed it into a main line running down the middle of the alfalfa field. We simply shut down the Waukesha, rolled the cross-pipes to the next mainline connector 200 feet away, reconnected and fired up the pump. The crosslines, one on either side of the main line were moved every two hours in the daytime. Tomorrow would be the last moving day – the alfalfa was almost ready to cut. The Waukesha ran on natural gas fed by a buried low-pressure line from the farmstead. By this time, I was hmm-ing almost non-stop.

Now it was back to Tom's through the wheat fields and rocking horses for a late supper. Tom said they always ate late in the summer-time. This made for a long day, I thought, and once again I was wrong. I only worked for these guys for a week but it was a culture shock every day!

After supper George brought his wife and baby over to meet me. Did he pull into the driveway in his 4x4? No way, Jose. This time it was another '59 Olds, a blue four-door sedan! By now I was humming like an old tube radio in warm-up phase.

Following a fine meal and some chit-chat I tootled off to bed with a full stomach and slept a sleep of the just and innocent. And what innocence – more culture shock awaited.

The next day after a good breakfast we hit the alfalfa field again. Dad had already fired up the Waukesha before breakfast. All day we moved pipes and did get-ready-for-haying-time stuff. At noon Mom had a light lunch ready. The day was hot but the house was air conditioned. I was already tired but ready to go when we arose from the table. I headed for the door, but what's this? They were going into the living room!

Well, glory be – it was siesta time! To escape the noon-time heat we all laid on a cool three-inch thick carpet, and desultory farm conversations morphed into snores. At 3 pm we got back to work. It wasn't such a long day after all.

That evening, on the ride back home with Tom, I thought about the 4-bys and the cars and said that farming must treat them well. Tom said it was tougher before they struck oil. I bolted upright – what did he mean, struck oil? Tom said all those pumps were on their land and they owned the oil rights! Those miles of wheat fields we passed were Van Pelt territory also! Now they rented out the wheat acreage and had fun with the horses and cows. I was ready to whimper and fall out the door – I had been sand-bagged.

Tom told me that when the oil money started to flow in late 1957 the first thing they did was to trade in their older pickups on the three new ones. This year it had been three new Oldsmobiles. I hadn't seen older brother's car yet – it was a loaded 98.

Working with Tom and George was more like fun than real work. We were busy all the time but they were always on my case. They thought my Canadian accent was hilarious – I told them theirs was no hell either, eh? On the Canadian prairies everyone curled in the wintertime. When I asked them if they did so, Tom said he never could learn to stand up on ice skates. I gave him a pass on that one –   I was not about to ridicule these kind folks.

One day we repaired fences. As we loaded fencing stuff into the pickup Dad opened a garage door and drove off in a 1948 Pontiac Star Chief straight-eight four door sedan, and that car was huge! It looked to me like a limo and later in the day I gave it a closer inspection. It had a tinted windshield, a hydromatic tranny and (get this) it had factory air! Now I'm an old car guy. I used to drive a 26-year-old Cadillac and I read old car mags, and never have I seen a write-up on a 48 Pontiac with factory air.

It made me wonder though. If this was Dad's daily driver then he was very thrifty or very old-fashioned. (he was both of those things.) I knew he didn't have a pick-up. Tom told me he had bought the Pontiac as a new car and it still looked new.

We ran into a neighbour on a shared line fence. We would ride the fence in our own pickups, each on our own side, and share repairs – but first it was chat time. The neighbour had his summer help with him, a high school teacher. Of course the attention was on me. I was an anomaly – what was a Canadian doing in Nebraska? I told him that I was out of school for the summer and gallivanting.

The teacher had been eyeing my six-foot-five-inch-ness making me a little bit nervous, and he asked me what grade was I in? Time for a white lie (rule 3.) I said I had been in grade 12 but wanted to go back one more year to raise my marks.

Now he wanted to know if I played basketball. Not very well, I told him. I said I was uncoordinated and clumsy. Things were getting uncomfortably dicey. It turned out that this guy was the high school basketball coach in North Platte. He said he could make me into a player. He said that last year they had been State runner-up. He said that with my height I would be the tallest player in the league. He said that I only had to stand under the basket and take a pass – and he was enthused!

And I was torn. All I could think of was what if I disappointed this guy like I had disappointed so many others. I had already figured out that I was an academic morning glory – bloom in the morning – fade in the afternoon.

I tried to get off the hook gracefully. I said I was in the country illegally and besides that I had no means of support while attending school.

This guy wouldn't give up – he had his bases covered. He said he knew people and the border thing could be fixed. He said the Alumni would bankroll me and with me on the team the State Championship was a lock. He was scaring the dickens out of me. This was way too fast, far too speedy! It was time to break rule 3 with an outright lie. I told him my parents had already enrolled me for September and had arranged for a tutor and the guy gave up.

The rest of the day went by quietly. Tom and George were unnaturally silent. As we drove back to the homestead they said they would see to it that I would have work for the rest of the summer. They wanted me to reconsider the offer and I almost did, but to tell the truth I was already getting a little homesick.

So it was would I, should I, could I? I'll never know.

On Saturday we built a windmill for a neighbour who wanted to add wind power to keep his stock tank filled. We all went over, including Mr. Van Pelt. Beside the well was a pile of steel stuff and around the well were four cement pads with bolts sticking up. They had been poured two weeks ago and were cured to cement perfection.

I thought this was sort of a help-thy-neighbour thing but it soon became clear that the senior Van Pelt was the ramrod of this outfit. The neighbour, a slightly younger version of Dad deferred to him, as did we.

At noon the older gents went to the house. Mrs. Kindly Neighbour brought us young fellers sandwiches and lemonade, and we sat on cool grass in the shade of a big tree. I just had to ask the question. Why, with black gold on their land, did Mr. Van Pelt build windmills? Tom said that in the depressed and dirty thirties there was always wind and the Ogallala aquifer always had water. If cash could be found it was spent on windmills to bring that precious stuff up to water gardens and stock. Dad had a windmill franchise, and selling and building windmills was what kept food on the table for three hungry children. Dad recalled those tough times only too well and would not or could not let the franchise go. Privately, Tom and George thought the windmill business was a pain in the butt, but they would never complain to Dad. To me all this was an indication of Mr. Van Pelt's solid character, and his boys were chips off the old block.

(And now I'm sure you've picked up on this, but big brother has been conspicuously absent from this narrative. That's because he turned out to be a supreme jerk.)

That afternoon just before the transmission and wind vanes were hoisted to the top

Mrs. Kindly Neighbour came boiling out of the house. "The cows are in the corn," she hollered!

Mr. Neighbour bolted for the barn to saddle up, I figured. Seconds later he emerged, not on but in his horse – a shiny light-green '46 to '47 Chev Stylemaster Coupe with not a speck of rust nor a dent, and off he went, bouncing across the pasture. Man, I loved that country!

Mr. Van Pelt pulled the windmill into gear. A light breeze caught the vanes and pure, fresh water poured into the stock tank. Mr. Neighbour wrote a check, shook everyone's hand and we packed up our tools and left. If it failed to rain next year Mr. and Mrs. Kindly Neighbour's cows would have unlimited water.

Sunday was a day of rest for the Van Pelts. We were going fishing. George and a friend picked me and Tom up and we went to the farmstead to hook up the boat – a sixteen-footer with a 25 horse Evinrude. Before we left, Mr. And Mrs. Van Pelt came out of the house dressed in their Sunday best. They headed for the garage and I watched out of the corner of my eye for the big Pontiac to emerge. Nope – the other door opened and the old folks left for church in an Eldorado hardtop – white over pink! Sandbagged again.!!

This day would produce yet another culture shock, and sort of a bummer. We were heading to a man-made lake on the North Platte River, and as we crossed into Colorado I mentally stroked a line through another state on my "Been There, Done That" list.

We reached the boat launch and I was disappointed. This was not my kind of lake – no shoreline spruce or jack pine and no rocky outcrop for little breeze-driven waves to wash. The whole area was rolling sandy hills and there was no wind today at all. This would be a hot day in Colorado.

They had a rod and a lure for me, and what a lure it was, and what would it lure? It was called a cowbell, an apt name if ever there was one. It was about eighteen inches long, starting out with huge curved silver thingies, which if attached to a hub could surely propel the boat. They reduced in size and were followed by coloured beads and tufts of what might have been buffalo hair, and ended in a hook which I doubted that a fish would ever see. It was supposed to be a fish killer. I opined it would surely kill a fish if you conked one on the head while casting. I thought this was funny, but the guys took it as an insult.

You could not actually cast the cowbell. If you caught a companion with that rigging you might knock him out of the boat, so you trolled – and trolling did not fit verb-wise either. With four cowbells in the water the Evinrude throttle had to be a notch higher. We <u>towed</u> the cowbells.

So we dragged the lake for a few hours. The other guys had a little luck – maybe a carp or sunfish. I had no luck at all. I could not tell if I had a fish on the line – all I could feel was cowbell. Just before noon I thought I had a hit and instinctively gave a little jerk to set the hook. Oh-oh – I broke the tip off my borrowed rod.

Now the guys were a bit disgusted with me. Canucks were supposed to be avid fishermen – what kind of a Canuck was this guy?

After lunch the other three went out alone while I rested in the sparse shade of a bush, wary of a snake attack and listening to the Olds radio. We went home to Nebraska – not a great fishing trip.

I didn't know it yet but Monday would be my last day of employment at the Van Pelts. The day as usual started off well. In the morning we were going to service the combine and grain trucks. The barley and oats were almost ready to be cut and in the afternoon we'd haul bales. Dad had already mowed, raked and baled the alfalfa while we were doing other things.

The grain trucks were in their sheds, but where was the combine? I soon found out. We drove down to what I had thought was the old derelict threshing machine and truck in the boneyard. It was culture shock deja vue all over again. What I thought was an old threshing machine was a pull-type combine. It was a huge, high animal, and not painted – all galvanized tin. High up beside the grain tank was a big motor of some sort. I was told that it was of the first generation combine design, purchased new in 1929 by a young Mr. Van Pelt. And what I thought was a junker truck was a 1929 Kissel,

also bought new before the crash. Of course it had the exterior old-age patina, but glass-wise, upholstery-wise and mechanically-wise it was in excellent condition. New pickups and cars notwithstanding, this family was frugal.

We serviced both units, put in batteries and gave them a test run. Then it was back to the house for lunch and a short siesta. Now it was time to haul square bales. George drove the shiny McCormick-Deering WD9 pulling the wagon. I was to be top-loader.

Now I found out what the Ferris Wheel dealie was. George drove the wagon alongside a bale. It entered the hamster wheel at the bottom and at the top an arm guided the bale onto a small platform. My mission (should I decide to accept it) was to place the bale in an appropriate spot on the rack. It seemed simple enough – grab a bale, stow it, repeat. The first 50 bales went okay – pretty darn steady, but I kept up. Then that rascal George gave the throttle a nudge. The alfalfa came up faster and I was like Lucille Ball on the chocolate assembly line. A full load was 150 bales, I was sweating like a pig, and at 120 bales I collapsed. From his tractor seat George saw the bales building a mountain so he shut 'er down and dug me out. I was a little woozy, so we unhooked the wheel and took the wagon load to the barn. While Tom helped George unload I was sent to the house for some water and a break.

When I went back out to the now-empty rack Tom and George were slightly apologetic. They should have known, they said, that a cold-weather Canuck couldn't take the kitchen heat and I should have had salt pills to avoid this severe sweat problem. Feeling apologetic didn't get me off the hook, though. Tom helped me with the next load and as we stored the bales I was still taking flack – I was now the goofy <u>cold</u> Canuck.

With the day's work done we ambled across the yard towards our respective vehicles. Older brother came striding out of his house with a piece of paper in his hand and a self-satisfied smirk on his face. Mr. Van Pelt was also joining us and he did <u>not</u> look very happy. Somewhere, off in the distance, a black cat yowled.

Big brother trotted up, waved the paper, and told me it was from the IRS. Illegal aliens could only earn 50 dollars in any one quarter. Anything over 50 dollars would not qualify as a legitimate expense, and furthermore, harbouring an illegal alien could result in charges being laid. Big Brother had been a busy boy last week. "You're done here," he said.

Tom and George were absolutely astounded. Dad looked unhappy – he must have been briefed by his eldest. It was strange – I was not upset at all. I knew the law was the law and I <u>had</u> jumped the border. The only thing that bothered me was that this jerk was enjoying himself.

Tom and George jumped in on my side. Border issues could be resolved – the family had some political connections. Wages could be deferred until a green card was issued. The family could sponsor me – a bond could be posted, but then Big Brother dealt the death blow. "This boy won't fight it. He's on the run from the law."

I was hot. I took a step towards Big Brother and he stepped back. Then I turned off the burners. I could not punch this guy in the nose in front of this fine family, and a punch in the nose would certainly bring flashing red lights.

Dad had seen enough. He went back to his house but before he turned away he shook my hand and said. "You've been a great help, Bob."

This was an undemonstrative man. He had always been polite to me but I thought he had hardly noticed my existence. He had given me a handshake! It was a medal of honour as far as I was concerned. I felt sorry for him as he walked away, an uncharacteristic slump in his posture. I knew then that he had given his eldest son financial control and he could not protect me at the expense of his own son's pride.

I had worked six days at eight dollars per. Jerkimer held out five ten dollar bills and I ignored him so Tom took the fifty bucks and handed it to me. I wouldn't take money from Jerkimer but from Tom I would. I had my revenge. I had spit in Jerkimer's face without opening my mouth.

That evening George came over and we had a beer and a chat – sort of a subdued going-away party. They still wanted me to fight – ignore the law remark. But the seed had been planted and for sure it was already sprouting in coffee shops and country kitchens. I thanked them, but I said it was time for me to go home.

The next morning Tom drove me to North Platte. I planned to go northwest towards Casper Wyoming and he took me to the edge of town. Once again there was a handshake, a "Good luck," and "Keep in touch." I said I would and watched sadly as Tom drove away – not a great start to my day.

That week at the Van Pelts in Nebraska made the whole summer worthwhile. I wrapped it, labelled it and stored it in the number one slot on my memory shelf. Number two slot already held the John Enns/Liberal, Kansas package. At the end of my odyssey the top shelf would be crammed. The second shelf, which held "interesting stuff" would also be full, and the bottom "bummer" slots would be almost empty.

Sidebar: Tom should have said "Keep in touch <u>soon</u>." Twenty years later I called Nebraska. There was no listing for Tom but I reached George. He seemed different now. He'd always been the most cheerful of the group but now he seemed more subdued – depressed even. I had to jog his memory and then he brightened up a little, He said Mom and Dad were gone, he was still on his own land, and Tom and Devoda now had a restaurant in Wheatland, Wyoming. He gave me their phone number, but he didn't mention Jerkimer. Hmmm.

Tom was glad to hear from me and we played catch-up. Yes, he remembered that summer very well. Yes, the restaurant was busy and he and Devoda were enjoying it. Yes, George was "hanging on" at the farm and of course, he also missed Mom and Dad. He didn't talk about Jerkimer either. On my end of the line I only mentioned what I had been up to. We mostly just talked about the good old days. I did not try to dig into the "Why are you in Wyoming" thing - I already knew why.

Here is my conspiracy theory and I am sticking to it. It was the "hanging on" that tipped me off. With Mom and Dad gone, the oil revenue drying up, and with Jerkimer in charge, he was after the whole shebang. Tom and Devoda were the first to go. George, as stubborn as he was cheerful in '59, was "hanging on." No wonder he sounded depressed.

I angled northwest and in this sparsely populated area rides were scarce. No worries mate, I had fifty dollars and 70 cents in my jeans. The scenery changed to rolling hills. This was cattle country and on the western horizon snow-capped peaks were slowly getting closer. My next to last ride of the day dropped me off at an intersection just south of Casper, Wyoming. It was rodeo day and they headed into town. I could have gone in with them but night was falling and I needed a hotel room and hotels would be full at rodeo time. Maybe I could find one farther west.

Casper was about a mile north, the rodeo grounds were on the south side and it was party time for sure. The evening was still and I could hear whoops and hollers coming from the bright lights and as dusk descended cars started to leave town – cars full of happy cowboys and cowgirls, all having a grand old time and totally disinterested in a hitchhiker. Then a cherry-top pulled up and the Sheriff got out.

He was perhaps 50 years old and all business. First, he got the rundown on why I was standing there and doing that. Then, for the first time since the border at North

Dakota my ID was checked. Now he also wanted to know my parents' names and my home address – I was getting the full treatment and I was getting nervous. The Sheriff picked up on that and did a 180. He patted me on the back and told me he didn't mean to be hasslin' me. "There's a lot of drinkin' goin' on tonight, son," he said. "There are good folks and bad folks, and I just want to know where to send your remains when I find your body in the ditch tomorrow morning."

He got back in his cruiser, did another 180 and went back to Casper. He was semi-alright after all.

Up to now I'd kept my hand in my pocket, but I had to get the H outta there. I stuck out my thumb – east, west or southbound – let's go, people! By golly, the next car stopped and they were turning west, and my last ride of the day would be quite a ride.

It was a '51 or '52 Pontiac four door, and I got into the back with my duffel bag. In the front seat were two good ole boys and a good ole girl and they were happy as larks. They told me that one of them had won the saddle-rasslin' event and I never did find out what saddle-rasslin' was. First prize had been a trophy and cash. They showed me the trophy and I knew where the cash was. It had been invested in bubblies, some of which they still had and did I want one? I didn't really, but just to be polite I popped a top and had a sip.

The guy didn't drive fast, but as often as not his half of the road straddled the white line and occasionally he used the left lane. Oncoming traffic was light, but every time headlights appeared I sat on the edge of my seat, hoping we would drift to the right. Finally I suggested that it might be a good idea if I drove. (Rule No. 6) The driver did not agree, but the other two passengers did, so I took the wheel. I ditched my can of beer, and they didn't notice.

They let me off at a county cross-road. They said there was a town ¼ mile north. If so, it must have been over a hill. This time it was me who said, "Good luck," and they yahooed south in a cloud of dust.

It was close to 11 pm – and it was so dark I couldn't see my watch. Stars shone overhead but there was no moon and I picked my way along the gravel road as my eyes adjusted to night vision. Already it was eerie. No dogs barked, and no night birds called – just silence. I got a whiff, just a hint, of the smell of old oil. The sun on its way across the Pacific Ocean had left a rim of light on the horizon. Outlined against that light I could see a few tops of what, at home, would have been old wooden forestry lookout towers, but I couldn't smell any trees and if there were trees one tower would suffice.

Now, just ahead and to the right I could pick out the town skyline (or more correctly, lowline.) I also saw a few low wattage orange/yellow lights here and there – about as bright as a coal oil lamp behind a window. Did this town go to bed early? Had they not paid the power bill?

I turned right and walked down the silent main street. The yellow lights were street lights, and they were pretty grimy – maybe the town crew was on holidays. I passed a few closed businesses before I got to the first street light. I was in front of a grocery store and through a large, dusty window I could see a counter and shelves – empty shelves. The next place was a barber shop and inside was a barber's chair with at least one more behind it in the shadows. A counter ran down one side, with barber stuff on top and a long mirror on the wall. The back of the shop faded to black and I couldn't see the end of the counter or mirror. What I could see was covered with dust – thick dust.

Well, Good Golly, Miss Molly! This was a ghost town – not some pretend tourist trap, but a real, honest-to-goodness ghost town!

I ambled along the block past a closed hardware store, a clothing emporium and a little house with a verandah and a faded picket fence. If there was grass in the little yard it would be dead grass. Across the street was what must have been a large lumber yard and beside it was an automobile dealership, with showroom windows so dirty they wouldn't reflect light if there was any. The last Buick had left the building years ago.

Ghost towns don't scare me a bit. I was enjoying this and could imagine it in its heyday. Cars and pickups would have been angle parked along the broad street and little bells over business doors would tinkle as folks went in and out. I could see children with ice cream cones and the barber shop full of guys getting haircuts and shaves, with old gents telling lies while waiting for a trim they didn't need. Behind me I could hear a roustabout trying to slam the door on his Dodge pickup. Heavy trucks rumbled by and horns tooted at pretty girls.

But I was getting edgy – where would I sleep? I was sure there were unlocked doors but I didn't want to spread my blanket on an inch of dust. Halfway down the next block I saw what might be salvation ahead. It looked like a two-story deal and a couple of windows on the main floor emitted a little light.

It was a hotel – old and faded, but it was open! I climbed a few steps and walked into a large lobby. It was dimly lit, but a guy stood behind the small counter. It looked like he was getting ready to close.

I said I needed a room for the night. He didn't say a word, just laid a little card on the counter-top. I printed my name, Canadian address, signed the bottom and he didn't even glance at it. He traded the card for an old-timey key. "Three bucks," he said. This was a man of few words.

"Quiet town," I said.

"Yup," he replied. "Oils gone." That's all. Just "Yup, oils gone."

He didn't strike me as being unfriendly. He just seemed to be tired and a little sad. He turned on the stairway lights and the upper hall lights. My room was the first one on the left. I found the pull string, and as the bare light bulb lit, the hall lights went out. As I closed my door I could hear him locking up downstairs. I had found a bed by the skin of my teeth.

I didn't expect much for three bucks but I was pleasantly surprised. The room was neat, clean and tidy. There was a high old-fashioned bed with clean sheets, pillow cases and a warm blanket. Folded at the foot was a puffy quilt, should the night turn cold. On the wall was a sink with clean towels on a rack. I would be able to wash and shave before I left. I crawled into bed, rolled into the centre valley and had a ghost-free sleep. I woke up at early dawn, wondering if I was locked in at the lobby. No problem – it was a snap lock and I pulled the door securely shut behind me.

Main Street in the morning light no longer looked romantic. This poor old town was just plain dead. There was no vegetation and everything was a dull gray like the dust on the barber chairs. Even the yellow rays of the beautiful sunrise could not cheer things up. As I walked toward the highway the fire towers started to appear out of the early morning fog. These were old wooden oil derricks as seen in James Dean's "Giant," and there were hundreds of them – north, south and west of town. All were weather-worn at the top with some hanging timbers and boards. The bottom half of each derrick had been preserved in oil, they would never rot. Most were only a hundred feet or less from their partners and over the rolling hills I could see more tops. Not one was bringing up oil, They were silent sentinels of past glory. As I stood at the road waiting for early morning passers-by, the fog lifted. The whole deal looked surreal, and now a shiver ran up and down my spine. Had I seen this last night I would have slept with my head under the covers.

Rides were scarce again. I must have nailed a couple of short ones, but before I reached Thermopolis another good guy picked me up. (We know that by now you're thinking that we are adjectively challenged – good people, fine men, kind housewives stir, shuffle, repeat: but you tell us; is there a better way to describe good, fine, kind folks without using four-letter words?)

This ride was a rancher in his '50s. We traded the usual info and I made a mistake, saying I was looking for work, which I wasn't, but I thought it would make me seem less bum-like. He asked me of I could ride a horse and when I said, "Not really," but I knew how to steer them, he laughed and said he couldn't help me out. Then we entered a construction zone and he said for sure there was work here. This guy was determined to land me a job. He stopped and asked the guy where to find the boss and the guy pointed to a little trailer up ahead. I tried to look as tiny and insignificant as possible in the corner of the pickup cab – not much luck with that. He pulled up at the trailer and he obviously expected me to jump out and hit up the foreman. When I hesitated he got out and said. "Come with me, son."

He walked into the trailer and without any chit-chat he said straight out, "This lad wants a job. Hire him." This rancher was what you might call proactive.

The foreman was startled, but he knew how to play his cards. He might need a man, what was my name, and did I have an SIN? I only had to answer the last question. No, I did not in fact have a Social Insurance Number. "Sorry, Can't hire you."

The rancher, of course, was disappointed, either in me or the system or maybe both. I told him I knew this would happen and I told him of the IRS rule. One long day on road construction would use up the better part of fifty bucks. It just didn't make sense for the foreman to go to the trouble. I said my best bet was to pick up a day or two on my road trip. He nodded in agreement, but this slightly gruff gent wasn't done with me yet. He was about to show me his true colours.

He was turning left to his ranch. At the corner was a restaurant and he pulled into the parking lot. "I'm buying breakfast," he pronounced.

I mildly protested that I had my own money, but that dog didn't hunt with this guy. We sat at a table and I ordered toast and coffee. He raised a shaggy eyebrow, and I told him I always ate light on the road. Once again he stated "Bring me a short stack, two sunny-side and sausages and bring this young lad the same."

Minutes later the waitress sat a huge platter in front of me with six perfect pancakes, swimming in butter and covered with syrup. Beside the flapjacks were two perfect eggs and the four sausages were not gherkin-sized like back home in Canada. They were positively huge and did they smell good! I was salivating like a hound dog and I dug in – light eater my ass! When I came up for air I glanced at my table mate. He was picking at his grub and not doing too good a job, at hiding a grin. Well, that old son-of-a-gun! I'd been sandbagged again! He hadn't been hungry at all – probably had a big breakfast at the ranch. He'd only ordered for himself to allow me to save face.

"Guess I wasn't too hungry," he said, "Here, have some of mine."

I was already stuffed, but just to be polite I took two more pancakes and a sausage. I leaned back, dropped my belt a notch and when the waitress came around with the coffee pot I asked her how high was a long stack.

"You don't wanna know," she replied.

When we parted company outside I added this guy to the fine, good, kind list – a bit sneaky also.

The next ride took me all the way to Thermopolis and I broke rule # 12. I was so full I had to have a nap.

Thermopolis was no bigger than its name on my road map, but by the time I carried my duffel bag a half mile or so I felt more alert. A few cars passed and a shiny blue '54 Pontiac two door came along with a lady driving. I never thumbed

lady drivers, figuring they would not stop anyway, and also, as a hold-over from my acne teen years, I never knew quite what to talk about. The lady pulled over on her own accord and picked me up. This would prove to be the strangest ride of the summer.

Up towards Cody the road climbed a bit farther into the low foothills, twisting and turning around pines and sandy knolls – beautiful country. Speed was not an option. The day was coming on hot and since she already had the windows open I stuck my right arm on the window frame, gave my hair a lick and leaned back - cool dude.

It started out innocently enough, trading ID, blood types – that sort of stuff. She was simply but neatly dressed – white blouse, dark blue skirt, no heels. She was neither pretty nor ugly – sort of plain, you might say. I judged her to be early to mid-thirties and she was pleasant to talk to. The car was plain but neat with no radio and a standard shift. I didn't ask her what she did for a living. This was a school-marm's car or maybe Marion the Librarian.

The conversation so far had just been about ordinary things, but as we ambled onward she seemed to shift gears. She started to talk about how women who picked up hitchhikers could be taken advantage of. There were bad men on the road – rapists, for instance. She started to get a little twitchy, sort of squirming in her seat and with every twitch and squirm her dress rode higher – nice legs! I started to get a little twitchy my own self. We drove through a little burg. This was her home town and I expected to be dropped off but she had other plans. She said she had nothing else to do and would take me a few miles further along. This was getting weirder with every passing mile. Cool dude no more was I. I was on sensory alert and sitting bolt upright.

She had started to calm down and so had I, when she played her last card, and it was a joker.

"I'm getting kind of tired," she said, "There's an old gravel pit just ahead. Let's go in and have a nap."

Now I've never been known to be the swiftest gazelle on the Serengeti but I'm telling you, my mind was racing. I could see into the future. Whether I said yes or no there was a 99% chance that this chick would holler "Rape!" I could feel the cuffs snapping onto my wrists. I was an illegal in a country with harsh rape laws and I'd be singing the Folsom Prison Blues for a hundred years. I wanted no part of this deal. Not me, babe – not now nor ever – not me, babe!

When she slowed to turn right on a little gravel road I grabbed my duffel bag from the back seat and I was out of the door while the Pontiac was still rolling. I bolted – and I never looked back.

I walked a couple of hundred feet north before I heard the car turn around and head slowly south. Now I looked back often, half expecting a deputy sheriff to come screaming around the corner – stress city. I picked a spot on top of a little rise with some bushes to hide in should a cherry-top Ford sedan come along. I wished the road behind was straighter – my back-sight was too short. An hour later I relaxed somewhat. The slowest Deputy Dog in the West should have shown up by now.

It was mid-day, the sun blazed overhead and I was dry. Off to the west the snow-capped peaks teased me. I wrote my name in the melting asphalt and if I stepped on the pavement my shoes stuck to the tar. A semi stopped and I gladly jumped in. On the dog-house sat a gallon thermos jug of ice water. He told me to help myself and I sure did, I surely did.

We chatted as the foothills petered out into flatter country. I didn't tell him about the weird lady, he wouldn't have believed it anyway. I hardly believed it myself.

This was a good ride. He dropped me off in downtown Billings, Montana. He had to go west with his load, and I had two miles to walk to the eastern edge of town – no big deal. I passed a laundromat, stopped and spent a couple of quarters and two nickels to wash some clothes. Further on I got a motel room, for seven bucks, I think. I was way ahead of the harvest now but I still had forty bucks and something would turn up.

My first ride out of Billings was a little weird. The guy was maybe in his late twenties and sort of hyper. The car was a VW bug, my first experience in a Volks. We headed north by northwest towards Great Falls and on to Sweetgrass. This was Montana, so the guy kept it on the pin and the miles flew by. He was a German, visiting his sister in Calgary who had emigrated to marry a Calgarian. It was her Volks and he had borrowed it to see the northwestern USA. He was a driving hand-waver and the bug had pretty much direct steering, so every time he waved the car twitchcd. I checked – no seat belts.

North of Great Falls he twitched, hit a pot-hole and a hubcap flew off, disappearing into the tullabies in the broad ditch. He stopped and we started to search. At the speed we were going the cap might still be rolling. We went back to the pothole and with me in the tullies casting like a coon hound, we worked our way north.

When the bug shrank to toy car size behind us I said it was time to give up. I told him to buy a cap from a salvage yard. His sister would never tell the difference. He said she was in love with her baby bug and would do an all-points inspection when he got home so I took a different path back through the ditch and found the cap. He tried to put it on but I told him it was probably bent and to leave it until he was a block away from sister's house. He dropped it in the back seat and said something in Deutschlander.

As we neared Sweetgrass he got twitchier – the border ahead was making him nervous. Had this been three or four years later I would have thought, "Hemp," but the deal was, he had no money left, all his cash was in the gas tank. Once back in Calgary he could wire home for money but right now he was worried that he might be refused entry at Coutts if they checked his wallet. He asked if I had any spare cash – he would return it in Alberta, so I loaned him two tens leaving me twenty dollars.

At Coutts the customs officer just checked his ID, passport and registration, easily believing that he had borrowed his sister's car. I was totally ignored, still under the radar behind my carbon-fibre shield. As we pulled out, back in Canada and heading for Calgary, I heaved a sigh of relief – legal once more.

The scavenger hunt notwithstanding, this had been a good, long ride and we would reach Calgary with lots of evening sun remaining. However, there was no offer of a twenty-dollar rebate, at least not yet. I bided my time.

In Calgary he was good enough to offer me a ride to the city limits – which way did I want to go? Up to now I'd given my itinerary little thought. Eastward was home and loved ones, but north led to the Peace River Triangle and I had relatives there. Peace River country was semi-famous, so Peace River it was. He took me to the highway north, just far enough to avoid urban traffic and it was pretty decent of him to do so.

But what about the twenty bucks? Decent guy or not, I was unwilling to pay cab fare. He had my twenty and a sister three miles away, but that twenty represented 50% of my stake and I had many miles to go. I gently reminded him of the deal and we had a short stare-down. He was five-foot-five and weighed 140 in a downpour. I was six-foot-five inches, 210 pounds, bats left and throws right. He blinked first and gave me my money back.

The next ride took me to Lacombe, just north of Red Deer. It was dark and I got a hotel room for five bucks. Hotels were as good as motels and were always cheaper back then.

The next day I headed for Edmonton and on northwest to the Peace River Valley. Once past Edmonton on the road to Peace the rides were longer. We were soon out of farm country and the broad highway ran through miles of uninterrupted, tall lodgepole pines. In pre-oil boom 1959, Whitecourt and Fox Creek were not much more than stagecoach way stations on sparkling-water river crossings.

I was dropped at Valleyview in the early afternoon. The ride was going to Grande Prairie, near my planned evening destination, but summer nights at this latitude are long so I headed north to Peace River.

The Peace River Triangle, as it was known in those days, was a slightly modified Pythagorean deal. One side led from Valleyview north to the town of Peace River. The hypotenuse ran southwest to Grande Prairie, and the triangle was closed by an east-west highway between Grande Prairie and Valleyview.

The scenery changed from pleasant to beautiful, with many valleys – deep, deeper and deepest. Some had small rivers at the bottom and others only little creeks. They certainly weren't large enough to cut these valleys but it must have been quite a torrent when the glaciers melted. On the high plateaux were fields of golden wheat, amber barley, flax and a lot of stuff called (brr) "rapeseed" (later to be rebranded as "canola.") Later that day I would see a big roadside sign: "Rycroft – Rape Capital of Canada."

Peace River was a nice looking town bisected by the Peace River itself. Going southwest on the triangle, I passed Bluesky – six grain elevators – five houses. (Many years later the Bluesky skyline would be featured on our Canadian five-dollar bill.) I passed Fairview, a fair-sized prosperous looking agricultural centre. (Also later, my oldest brother Walter would live and work in Fairview, first as a high school principal before becoming superintendent of the largest school district in Alberta.)

Not too far south of Fairview I crossed the Mighty Peace again at Dunvegan and it was a sight to behold – the deepest valley I had ever seen. The highway just seemed to drop into space before anging down to the ferry crossing far below. At the top was a warning sign: "Trucks – Select a Lower Gear." Well I guess so!! Cars take heed also!

It was at least two miles down to the ferry. The river was broad and mighty for sure, but it seemed I was in a different world and I actually was. Dunvegan has its own micro-climate and truck farms still flourish in the valley at Dunvegan.

Sidebar: I would cross at Dunvegan many times twenty years later, now on a long, high bridge. The oil boom was in high gear and I was hauling oil patch cement. The hills on both sides were still pretty long and potentially dangerous, but every time I crossed I got a glimpse of the old ferry crossing a half mile below and remembered that young me and my innocent wonderment.

My ride climbed slowly out of the valley. The truck ahead could not get out of his lower gear until he reached the south plateau. At Rycroft, not far from Grande Prairie I said goodbye. I was going to call Aunt Leila.

There was a phone booth outside a restaurant but Aunt Leila and Uncle Irwin were not listed. A couple walking past had never heard of them but these folks were new in the area. No problem – Uncle Irwin's nephew owned a Chrysler dealership in Grande Prairie. I would chug on down and contact my second cousin once removed. He would know where Leila and Irwin were at.

Table Scrap

Leila and Irwin had sold out in Saskatchewan years before, moving to a homestead at Rycroft. For quite a while now they had been stopping every fall at my parents' place in the Rainy River Valley on their way to Florida, driving last year's Chrysler. From Fort Frances they would head for Windsor, Ontario to pick up this year's Chrysler. Last year's would be dropped off at a Windsor dealership. Grande Prairie had already established the trade-in value. The Windsor guy got the car at wholesale and could easily demand top dollar for a one-year-old salt-free set of Alberta wheels. So you might say that farming in the Peace River Triangle was hand-to mouth – or Chrysler-to-Chrysler.

Irwin told us that one year as he was dumping the last truckload into a grain bin, Leila was warming up the already packed car. Happiness was driving across Saskatchewan with the first winter blizzard in their rear view mirror.

In Grande Prairie I went into the lobby of an elegant hotel and called Blacktop Motors (name changed to protect the guilty.) I called the owner, told him who I was and asked how I could contact Aunt and Uncle. He said unfortunately they had sold the Rycroft farm and had retired to Saskatoon, Saskatchewan. However, he sounded pretty pumped. Great Aunt Maggie May lived on their ancestral farm just out of town. She was in her 90's, was the family historian, and would be delighted to meet me. He would pick me up after the dealership closed.

All right! I got a room – twelve dollars, and it was worth it. It was the last room available, the prime-ministerial suite. I cleaned up, changed all my clothes and gave my loafers a spit and wipe.

When the guy walked into the lobby I immediately knew the game had changed. He looked unhappy and had his sourpuss face on. He introduced himself, as did I, sticking out my hand. He barely brushed it, pulling his hand back quickly to avoid any contamination. Holy cow! Had Nebraska Jerkimer put out an APB? My once removed second cousin had moved me down to the fourth ladder rung.

He said Maggie didn't feel too well and maybe it wouldn't be a good idea to visit her. It sounded like a flimsy excuse to me. His body English said it all – he thought I was a bum. Then the jerk held out a twenty-dollar bill holding it just at the narrow edge between thumb and forefinger. The bugger was buying me off!

My synapses were firing on all cylinders, once again driving me to the head of the herd. My inner financial adviser brought my personal account up to date – twelve bucks and change. If this jerk was ignorant enough to offer me twenty I could be a jerk also. I broke rule #5, took the money and he weaseled out the front door. Perhaps a good meal would make me feel better, so I went into the dining room and had the pork chop special. It was tasty and I was happy to trade in the ugly twenty for cleaner bills, but I still went to bed feeling like the bum I wasn't.

On the road again at daylight, my first ride was the usual nice, kind, good people and when I completed the triangle at Valleyview I was back in the groove, enjoying the good times and rolling with the bummer punches.

Back to Edmonton took about as long as it takes to write this – one ride, and dad was driving. Eastward on the Yellowhead was also easy. I don't recall any specifics other than one guy, who, proud of his province, gave me an Alberta tutorial.

He said Alberta was cattle and cowboys and cash crops. The Calgary Stampede was the largest rodeo in the world. Alberta was potatoes and sugar beets at Taber, dinosaurs at Drumheller, tourism at Banff and timber to the north.

Alberta had oil – lots and lots of oil. In 1947, Leduc #1 in the Drayton Valley blew in and in 12 short years Alberta oil production eclipsed the total produced by the now tapped-out field at Petrolia near Sarnia in Southern Ontario.

And he said Alberta was rat-free! No kidding, I didn't know that. Yup, rat-free. There was a rat fence on the Alberta-Saskatchewan border patrolled by rat police carrying 220 Swifts. There was no appeal process for border jumpers – just a varmint bullet in the head. No rats allowed in Alberta.

This was a point to ponder and ponder I did. He had not mentioned Montana rats and I had not seen a rat fence at Sweetgrass/Coutts. How about Northern Alberta? When rat country changes to muskrat country does the rat fence continue on until it hits the Northwest Territories? Were there only Saskatchewan muskrats?

I pondered along eastward on the Yellowhead eventually filing the rat question in with other important stuff such as: why is it that when you boil an egg it gets hard, but when you boil a potato it gets soft?

Later we passed through Lloydminster into Saskatchewan. Lloydminster sits on the border, bisected by the Alberta-Saskatchewan boundary line, down the middle of its Main Street. There was no rat security gate and no rat watchtower with searchlights or varmint guns at the ready. Alberta Homeland Security was lacking in Lloydminster.

Table Scrap: On Muskrats and Potatoes

Four years later I spent a summer looking for gold 230 miles north of Yellowknife. At base camp I met a young fellow from Stoney Rapids, a native community right on the border where Northern Saskatchewan meets the Northwest Territories. The lad was sixteen, had grown up in that isolated community and this was his first job. He had come to Contwoyto with his father, a diamond driller, and had been hired on as a cook's helper. He was a nice, friendly energetic boy and a good worker.

I was told by a cook that when the kid was given a potato to peel, he had no idea what to do with it – he'd never seen a potato. I didn't believe that one, but I did believe a story that Barney (the kid) told me himself.

His uncle was a good trapper. Legend had it that in 1947, with pre-Bridgette Bardot fur prices high, he had brought out his winter's catch to the fur sale in Prince Albert and along with the high end mixture of lynx, wolverine, beaver, fox, marten, fisher and the like, were four thousand muskrat pelts. I had no trouble at all believing that – muskrats are a trapper's late winter bread and butter. They are easy to trap and skin and in 1947 were probably worth five bucks each. Four thousand might seem to be an exaggeration, but muskrats, like ordinary rats, are prolific breeders. It's hard to trap out a muskrat patch and good trappers don't decimate, they harvest.

Anyway, Uncle got 40 grand for his fur, also not hard to believe. What I surely did believe was that Unc never got out of PA with his forty thousand dollars. He returned to Stoney, stony broke. (Trappers are like diamond drillers and prospectors – hard come, easy go.)

Table Scrap: Another Stoney Story

This was also told to me at evening tea-time while in the NWT and I have no reason to doubt it's veracity.

They tell me there is a large anomalous magnetic zone east of Great Slave Lake, where a conventional aircraft compass cannot be trusted.

A young, relatively inexperienced bush pilot is heading to Yellowknife (from Baker Lake?) navigating by compass and probably trusting it more than he should. By the time he realizes he has been led astray, he is off his map and lost. Short on gas now, and out of contact with his base, he lands on a lake not too far north of Stoney Rapids and makes two fateful errors.

It is coming spring and he doesn't want the Cessna to sink in the lake when the ice melts, so he ties it up on a well-treed shore where he will wait to be rescued, but he is well off his planned flight line.

A search is initiated and broadened, but even if someone had flown over, the Cessna is hidden by trees. The search is eventually called off.

Later that summer some fishermen spot the plane. In it is a note: "Waited three days – I am walking west."

Had he walked east across the ice, within a mile he would have picked up a well-travelled dog sled trail leading south 50 miles to Stoney.

He was never seen again.

I reached Saskatoon just before suppertime. I called Aunt Leila, and Uncle Irwin picked me up with their current Chrysler. They were so glad to see me. Right back at ya, Aunt and Uncle, and supper was already on the table in their comfy retirement home. Aunt Leila was apologetic about eating leftovers. I dug in like a starving puppy dog.

I spent two days there, and they couldn't do enough for me – big breakfasts and lunches and roast beef and mashed potatoes and gravy for supper. I held up my end – 80% for me, 20% for them.

Aunt Leila wanted to wash my clothes. I drew the line there, as I was only one day away from my sister in Winnipeg. I did have a shower and when I looked in the mirror I saw that my right arm was now brown and my left was white! The Roswell alien was also a shape-shifter!

Uncle Irwin gave me a tour of Saskatoon. Not much to see other than bridges. The South Saskatchewan runs through the city and we must have crossed the river ten times. Saskatoon has more bridges than people!

In the evenings we talked about family, new days and old days, with the volume low on the ignored TV. On the third morning Leila and Irwin took me back to the Yellowhead. When I pulled out my duffel bag, Uncle Irwin got out, came around the car and motioned me to the back bumper. The old rascal stuck a ten-dollar bill in my shirt pocket. We shook hands, I gave Aunt Leila a smooch through the car window and I walked down the shoulder of the Yellowhead in an aura of family love. I would never see Leila and Irwin again.

I don't recall any of the rides that day, only that lifts were plentiful and I arrived at the west edge of Winnipeg about 6 pm. I called my sister, and my brother-in-law and two young nephews picked me up. I wasn't home yet, but at Beth and Leon's I always felt at home. Beth gave me a hug and a smack on the side of the head. I deserved both.

I spent a couple of days there playing with my nephews while my sister washed my duds. I filled them in on my summer, and while I think they may have had some doubts regarding my truthfulness, for sure they had doubts about my sanity.

One day Leon drove me to the tire shop downtown. Call it stubbornness or big cojones, but they owed me money and I was going to get me some. The manager didn't like it, but I pointed out that I <u>had</u> sold tires. He reluctantly stroked me a check for thirty-five dollars.

I went back to Kenora and south on Highway 71 to the farm, where I was greeted warmly by my mom and a little less so by my dad. I spent a few days at home helping out a bit and repacking stuff. Mom was in on my plans and made sure I had everything I needed in my dad's old Gladstone bag. I would be hauling two pieces of carry-on luggage this time.

I pulled out on Sunday, Labour Day weekend. Dad or little brother may have driven me the half-mile to the highway. Dad wondered why I wanted to go back to Trenton, so I told him a big white lie – I told him I had a job lined up.

(I had a job lined up? Right. For the last sixty days my plans were more of the "When I go to sleep, I'll probably wake up in the same place," variety. Actually I did plan all along to finish up at Trenton. I had to complete the circuit.)

Up to the Trans-Canada was fairly good going. Heading east was a bit slower. It was the last long weekend of the summer, and most people who were going somewhere were already there. The only memorable ride was a doozy. Two guys in a stupendous black-over-red brand new Impala four-door 360-degree glass hardtop –

all gleaming chrome and with that new car aroma. The two guys were brothers and joint owners of this beauty. They had just come off the Dew Line and had paid cash for the car. As usual, I was envious.

I caught the ride at Ignace and they took me all the way to the Lakehead and were good enough to drop me at the east edge of Port Arthur before turning back. I got to Nipigon before dark and luckily (on a long weekend) got a cheap motel room.

The next morning at 6am sunup I walked across the Nipigon River bridge to the junction of 11/17. In 1959 one had to take the north Trans-Canada. The last gap in the Highway 17 (southern) route on the north shore of Lake Superior would be filled in next year.

Slow again! It took me five hours to get a few miles past Geraldton and I was dropped off in the middle of nowhere. My ride was heading north to his bush camp on a seldom-used side road.

From the east appeared a black-and-white. I wasn't concerned – cruisers often passed me by. But this guy stopped, backed into the side road and crooked a finger at me. I walked over, heading for his side expecting a short chat, but he pointed to the shotgun seat and I got in.

This young officer was all business and right from the get-go I felt he had me scoped as a hardened criminal. He wanted all my ID including my birth certificate. He made me show him my cash reserves and family snap-shots! (Fortunately, I had already ditched my bogus liquor permit two months ago before entering the USA.) After making me sweat he pointed to the door – out! He pulled out, turned east (the same way I was going) and disappeared around a corner. I was glad he had not offered me a ride.

I walked back to my two-piece unmatched luggage set and ruminated. The old sheriff in South Dakota had been so kind, the young Hoooouston State Trooper had been very polite and the Casper sheriff had turned out to be sort of fatherly. I had to come all the way back to Ontario to be treated like dirt, and I've often wondered if that cop had long, miserable career.

(Just a couple of years ago my retired cop friend told me that around the Ontario Provincial Police water coolers, Hearst was known as the "penalty box." If an officer had one or more black marks for a myriad of non-firing infractions he was sent to Hearst, and nobody liked Hearst. Perhaps grumpy-cop was stationed in Hearst, doing two years for (not) cross-checking.)

The next ride took me forty miles to Longlac and let me tell you – Longlac was something else! (Warning: the following stuff actually did happen. Only <u>my</u> name has been changed to protect the guilty.)

I was let off in Longlac just before noon. I stood beside the road waiting for a ride to come along – and I stood, and I stood, and I sat on my suitcase, and I stood. Nothing was moving in Longlac. The air was still and rain clouds threatened, but even they stood still. It may have been raining somewhere bit I guess it don't rain in Longlac.

Off to the southwest I could see a sawmill, shut down for the weekend. On the other side of the highway, across from my "FREE TAXI" stand stood a one-bay, two-pump service station with a hand-lettered sign in the window "Closed for the Weekend." The sign should have been on the edge of town.

A small residential area was a short way north, separated from the Trans-Canada by a gravelled street. A car went by on that street followed by some lazy dust. A screen door slammed somewhere and farther off a dog found the energy to bark twice. There was no business district that I could see other than an old two-story insul-brick sided hotel a few hundred yards back. It stood just off the highway on another angled street – no action there either.

Two semis passed by, heading west. In those days, at least in Ontario, trucks could not run on Sunday unless hauling perishables. No trucks passed me going east. I guess today no one was about to perish west-east speaking.

A car came out on the highway and I thumbed it. I was ignored and the car turned off shortly after. I thumbed a couple more, with the same result, and thereafter I kept my thumb to myself. These must have been church-goers.

Not one car or pickup came from the distance to disappear in the other direction – not one. It was Sunday morning coming down in Longlac. Oh my – stuck in Lodi again.

I stayed there all afternoon, wishing I had a book to read. I was hungry and thirsty and I had eaten the last of my mom's sandwiches. It was futile to walk east toward Hearst. Remember? "No Gas for the Next 120 Miles."

Just before dark, a plain-jane '59 Chev pulled onto the highway and drove by slowly with three guys and three girls around my age having fun. It was obviously a local, and the car was full, so I just smiled and waved. Then they pulled over two hundred feet ahead and the driver stuck his arm out the window and waved the 'c'mon" signal.

Any other day I would not have taken the bait, but I had been waiting over six hours for a ride. I grabbed my bags, trotted to the car and just as I reached for the rear door handle the Chevy dug out, spitting gravel. It departed leaving loud laughter and yells of "Sucker!" in its wake.

Was I teed off! Embarrassment at my own stupidity turned to anger and fifteen minutes later the car went by going west. The crowd was still jeering so I broke rule #7 and broke is not the word for it – I shattered rule #7. I walked halfway to the centreline and gave them a middle finger salute. No half-hearted finger-flick either – I made darn sure they saw my estimation of their collective IQ number! The jeering stopped and the car turned off into the village.

I had fully expected them to turn around, come back and give me a thumping (if they could.) In my present mood I even welcomed some action. Those three guys looked a bit scrawny, but I was a road warrior. I had been packing on muscle all summer. I thought they might have been going for reinforcements, but within an hour I relaxed. They probably couldn't find anyone in this burg with the gumption to crawl out of their La-Z-Boys to help them deal with me.

It was now as black and silent as an abandoned coal mine, and even if a car or truck did come by there was as good a chance of being run down as being picked up, so at 9 pm I packed it in. Few lights were visible downtown, but behind me a couple of lit windows at the hotel beckoned. The owners' suite, perhaps?

I entered a tiny lobby – one armchair and a small reception counter with pigeon-holes on the wall behind it. A closed door was on my right, a narrow staircase on my left and a single bare light bulb in the ceiling – no valet parking – no bell-hop. On the counter was a little silver bell. I gave it a ding – silence. I gave it two more cracks – no action. From somewhere in the rear I heard the faint sound of a television set. I gave the bell six hard blows and still there was no response, so I considered what to do next. Should I hammer on the door? I knew not where it led. Should I open the door and holler? That seemed to be kind of cheeky and I didn't want to invade anyone's privacy, so I decided to be my own desk clerk.

I spun the wheel and randomly selected a pigeon-hole. I took out the registration card, filled in my particulars and traded the card for the room key. Upstairs I found lucky #7 not far down the hall. It was a simple basic room and I had a simple basic night's sleep. I would settle the room account in the morning.

At 6 am I was ringing the bell again, and again no one stirred. No cars were parked in front and with no guests there would be no need to open the dining room even if they had one. The owners were also having a long weekend.

Had I a five-dollar bill I would have left it on the counter with my registration card, but my smallest bill was a tenner and the room wasn't worth ten bucks. I put the key back into its slot, took the card and walked out the door.

(Over the passing years Longlac lays softer on my mind. Forget the long, ride-less day and the local punks. I only remember this – my cheapest room ever!)

Six o'clock in the morning is kind of early to start out on a holiday. It took a while for the first vehicle to pass and I was mildly stressed out – ready for Papa Bear to come boiling out of the hotel looking for "Who has been sleeping in my bed?"

Then I formed a calming-down theory and here it is: The hotel survived on its beer parlour, not on room rentals. The only time a room would be needed was if some bushwhacker stayed too long at the bar. The used bed might not be discovered for days – weeks, even. When it was discovered, Mr./Mrs. Owner would chew out the cleaning lady for missing it last February. I was in the clear.

An hour later a semi appeared from the west pulling an enclosed furniture van. The tractor was a Ford cab-over and I stuck out my thumb. The driver gave me an apologetic shrug and pointed to a pile of furniture blankets on the passenger seat. I waved and smiled. As he passed I saw lettered on the trailer "North American Van Lines." On the tractor door it said, "Jet Moving and Storage, Winnipeg, Man." I correctly assumed he had left Winnipeg at dusk and had driven through the night.

The next vehicle was a '57 Oldsmobile station wagon – a land yacht based on the four-door hardtop.

In the front were two adults and in the back seat, two children and a dog. Way in the back I could see a pile of luggage and camping stuff. I didn't thumb them but they pulled over! Mom asked <u>me</u> if I needed a ride! I sure did and Dad got out, and rearranged the back to find room for my stuff. I got in with the kids and saw that Mom was visibly pregnant. A prototypical nuclear family – 2.5 children and a dog.

Had this been the only ride of the day I would have been satisfied, but by day's end my faith in people was once again totally restored. Yesterday the pendulum of life had hit the left-hand stop. Today it swung back to the right and would stay there for years to come. All those "good lucks" and parting handshakes were still in the bank and earning compound interest.

They took me the 120 gasless miles to Hearst, chatting all the way. The dog didn't talk much – he liked having his ears scratched. We passed the Jet Moving and Storage truck.

At Hearst they let me out with apologies. They were on their way to Toronto but they had some people to visit, and should they catch up to me I would once again be joining them. No apologies necessary, and they drove into town leaving me with another "good luck" in the bank.

The furniture van passed. I waved and smiled and so did he.

The next ride took me to Kapuskasing. We passed the van.

At Kap the van passed me – by this time we were both laughing.

I got a ride to Cochrane where #11 turns south. I had passed the van again!

Fifteen minutes later, here comes the van and I held my hands up in the "It's out of my control" position and the guy stopped! He stowed the furniture blankets in the sleeper along with my stuff. I climbed in and he told me he could just not pass me by again. We were already old friends.

As he ran the gas job up through the gears he asked me where I was headed and I told him I hoped to get to Trenton today if I could. He said I'd probably make it – his load of furniture was going there. Bonus! Double bonus! (And this would be a triple bonus day.) We passed through Cobalt and rounded the infamous Cobalt Corner.

Sidebar: Ontario's Kings Hwy 11 is called the longest Main Street in the world. Starting in Toronto at the south end of Yonge Street on the shore of Lake Ontario, it makes its way northward and westward, ending 1200 miles later at the International boundary one mile west of the town of Rainy River. The last gap between Fort Frances and Atikokan was completed in 1964. Number 11 now bypasses most towns, but it is still known as the longest Main Street in the world.

(I have travelled Hwy 11 many, many times and I now live two short blocks from mile 1198.)

In 1959 the highway still ran through Cobalt and in Cobalt was the Cobalt Corner. At that time most of the town was built on a south-easterly slope and partway down this slope Main Street had a 90-degree corner. A big house stood high above the northwest corner and in order to have a useable yard, the owners had built a twelve-foot high cement retaining wall on their east and south lot boundaries. Main streets were never wide in Ontario but at that corner there was no room for sidewalks and on the downhill side there was hardly any shoulder to speak of.

The wall made it a blind corner. Cars and single-axle trucks had no problems here but semis did. On my way north in the spring I had noticed in passing that the retaining wall had taken a beating over the years.

As we drove and chatted it up, I asked the guy if he had ever been through Cobalt – he had not. Up to now his hauls had been west of Winnipeg, so I filled him in.

At the corner I acted as flagman, standing on the far side of the corner. When I saw a gap in the traffic I waved him forward. He had to approach the corner with his driver's side wheels on the opposite shoulder before cranking the steering wheel right. I held up oncoming vehicles while he completed the job with inches to spare. I jumped back in the cab and we were on the road again with no scratches or missing concrete to account for. Semis needed a co-pilot. I should have set up a Cobalt Corner Consultant Corp. (The highway now bypasses Cobalt and the only sign of Cobalt's existence is a sign pointing east.)

We talked up a storm as we motored onward. The guy was a transplanted German and had been in Canada with his wife and child for three years. He lived in Winnipeg and had been sponsored by the Jet owner. My host had started out with a broom and had graduated to roustabout and on to city deliveries. He had now been on the long haul for six months. He must have learned his English years before or else he was one sharp-eared cookie – his speech had only a trace of a Teutonic accent.

As for me – I had only my trip to brag on, yet he was very interested and impressed. He had seldom seen hitchhikers in Germany. Backpacking was starting to get popular over there but that involved hiking without the hitch.

We had lunch and gassed up the Ford at a truck stop on a hill high above North Bay. We drove down another Main Street and continued south.

He didn't show it – a little reddish eyeball-wise – but he must have been getting tired. He had now been on the road for nearly twenty hours. I did my job well, keeping up both sides of the conversation. I was nearly at journey's end and I was alert enough for both of us.

We pulled into Trenton around 8 pm and he offered to drop me off downtown. I had a better idea – I would help him off-load and he accepted my counter-offer.

Three hours later we were back downtown. He offered to pay me which I politely declined, so he bought me a beef and barley sandwich in the bar of the Gilbert Arms. (And the Gilbert Arms did have a bellhop.) As we sipped our apres-sandwich beer, he, impressed by my willingness to pitch in, gave me some very valuable information – info which would lead to a job.

He told me a big Air Force move was underway. All across Canada, van lines were criss-crossing thousands of miles shuffling married personnel and their families. His affiliate was in Belleville. Fox Cartage was the local Allied Van Lines affiliate, and I should check it out.

We said goodbye with a promise to keep in touch (which I kept five months later.) He went on to Belleville where he would sleep in the Ford's bunk. I got a room at the Gilbert Arms. It was expensive but I deserved it and it would be my last hotel room for some time.

Before hitting the hay I pulled out my trusty road map and did some figuring using all ten fingers of my hand-held calculator. In the last two months I had travelled over ten thousand miles on less than $200. I had graduated Fiddle Foot Cum Laude – only some post-grad courses remained.

And, I would never, ever, hitchhike again!

CHAPTER VIII

Fiddle Foot Training - Post Grad Sept '69 – June '70

I awoke the next morning virtually penniless and with three things on my want list. I needed a place to hang my hat, I needed a job, and I needed wheels.

The first part was easy. My cousin Patsy was still in Trenton and worked just across the street. During her lunch break we walked to her landlady's place nearby and the old girl agreed to let me bunk out on the sofa until I found better accommodation. She also agreed to wait for compensation! People tended to be charmed and bewitched when Patsy was around.

Table Scrap

Back in June on a warm afternoon just before my thumb departure from Trenton, I borrowed my brother's car and Patsy and I went to a fine beach near Carrying Place. We spread a blanket and swam in the warm water, and we had the whole beach to ourselves – but not for long. An Air Force Single Otter on floats landed, beached nearby and five or six young Air Force types got out, unloaded some gear and pretended to do Air Force stuff. They were a Search and Rescue team on manoeuvers, which meant that mostly they were manoeuvering as close as possible to Patsy. It was fun to watch those young guys. Of course it was too hot for heavy shirts and it soon looked like Muscle Beach and the lads kept tripping over stuff. It was tough for them to keep an eye on Patsy and their own feet at the same time.

A couple of months earlier Patsy had taken me to a Sergeants Mess dance at the Air Force Base. It was held in a huge hanger and attended by a huge crowd. I sipped a 30 cent glass of draught beer while a line of at least fifteen young uniforms took turns whirling Patsy on the dance floor. Pretty Patsy could sure draw a crowd.

Next on the list was a job. I walked four blocks east to the Fox Cartage compound. It was a warehouse with a loading dock at the rear. I walked into a spacious room with a bank teller's cage at the far wall and behind the iron bars sat a young lady. "Tough place," I thought.

The young lady didn't look tough – more of an "I've seen it all," attitude. I told her I was looking for a job. She said there was nothing this late in the day – come back tomorrow. She wrote down my personal info. No phone, no address – name only – that's all she wrote. And she wrote it herself – my illiterate "X" would not do.

As I walked back uptown I had only a few misgivings. The lady's attitude was not encouraging, but the Jet guy had seemed to know what he was talking about. I didn't need a car yet. I could walk the six blocks to Fox, but I had some time to kill, so I stopped in at the pool hall.

There was only one guy there. I had known him last winter and had received some pea-pool tutorials at Pool Hall U. He was now on his own career path, moving up a notch to semi-pro. He asked me if I wanted to shoot a game. I still had some loose change in my pocket – why not?

As I learned another pool shark lesson I mentioned that I might be in the market for a car. Well – Dalton had a car to sell – why not call him?

I'd known Dalton last winter also and I knew the car. It was a 1949 Merc four door, a bit shabby with some bondo patches, but Dalton had driven it to school every day, so I called Dalton from the pool hall.

Sure, he'd sell me the car, and he came down to pick me up, but not in the Merc – he had upgraded. Before I saw the car I told him I'd buy it, but I had to wait for my first payday. Dalton said he would wait. I could take the car and pay him the $75.00 when I had it. He gave me <u>one</u> caveat emptor before I left – the Merc would only start when cold. I told him that was no problem at all. I would only need it to drive to work and back home again.

So now I had lodging, transportation and maybe a job. My credit rating has never been that good since.

Customer: "I like that red car."
Salesman: "You don't want that car – it has no reverse gear."
Customer: "That's OK, I'm not coming back."

I was back at Fox Cartage at 8 am the next morning. I walked in the door and eight guys had beat me to it. It was a dockyard shape-up deal – first in, first out. A trucker would walk in and have a short chat with the Iron Bar Lady and she'd point a finger at one or two stevedores who would then leave with the driver. She was a girl of few words and I didn't get the finger until 10 am. Tomorrow I would come in earlier.

I knew what my job was. I was grunt labour – no brains required – unload and turn that rig around. This reshuffle would not last forever and every load missed was one a driver would never get. So I trotted up the gang plank, into the cargo hold and trotted back into the house with cartons and furniture. If the lady of the house was present she would point out my destination. If not, no sweat: cartons were marked – kitchen, living room, bedroom – and I trotted. The first van was off-loaded lickety split and although the guy's back haul was in Belleville, he was happy to have a quick turn-around and took me back to Fox. Another truck was waiting for a horse and I was off to the races.

I learned the finer points of furniture moving as the days went by. Some drivers were friendly and others were not, but I cut the grouches some slack. How many hours had they driven and when would they get home again? I worked steadily and silently and invariably, when the van was empty the grouchiness was gone. The wages were $2.50/hr with no overtime, but I found out at the end of the first week that some drivers had padded my hours – bonus!

I also learned, when I got my first pay envelope, that I was once again under the radar – cash only, with no deductions. I paid for the Merc and paid Patsy's landlady, thanking her and telling her I'd found a place to stay. She was happy to see me leave – my hours were erratic and I had more than once woken her up in the wee hours to let me in. I explained this to my new landlady and she gave me a key.

I was continuing my human nature studies and some of the things I learned made me feel bad. All my fellow stevedores were older than me. Most of them were in their late twenties or early thirties, picking up some extra cash for a new boat or a planned backyard patio. It was the guys in their forties I felt bad for – they were generally alcoholics. The Missus had control of the Air Force paycheck and they were earning cash for their next bottle. They were all nice guys and tried hard, but they tended to power out.

It was a good cardio workout every day. My legs had been toughened last summer, I was adding upper body muscle and I was gaining a reputation. Sometimes a familiar driver would walk in and point to me and once I worked 36 hours without a break. One payday, when sliding my brown pay packet through the little gap below the grillwork, the Iron Bar Lady actually smiled!

One offload I will always remember. It went to an upscale neighbourhood in Belleville. It had to be an upper echelon officer move – the house was upscale, the furniture upscale, and the fortyish Mrs. Officer was definitely upscale, if you catch my drift.

Right off the bat I got into trouble. While trying to manhandle a heavy chest of drawers up the stairway, I gouged the freshly-painted drywall, and was she ever mad! Back in the van the driver told me not to sweat it – the company would repair the damage but he told me to slow down and be a little more careful.

So I did, and I had time to observe and cogitate. Mrs. Officer would not even look in my direction – she was much more interested in the young, handsome driver. Before we finished that load I expected her to offer to help him carry in the bed. I mentally filed this under "Possible Future Career Path." Big-bosomed waters run deep.

Table Scrap

I had already formed a theory on how to use Miss Mercury's no-start-when-hot habit to my advantage. Would it work better than the we-are-out-of-gas gambit? So on one of my more quiet evenings I asked a girl for a date – I'd known her at school and I liked her.

We parked to watch the full moon light up the Bay of Quinte and smooth-dude made his move.

She said she had to go home. I said the car wouldn't start for at least one hour. She said to try it anyway, and wouldn't you know it – Miss Mercury fired up.

We drove home in silence. One reason I liked this girl was because she was smart – too smart for Suave Bob.

The Air Force move was slowing down. I needed a new job and a better car. For weeks I had been passing a used car lot and had been admiring a metallic brown '47 Plymouth coupe. I was pretty stakey now, so I stopped in and made a deal – my car and X dollars and I drove away as fast as I could before they tried to start Miss Mercury. I actually felt guilty for almost 24 hours, until my muffler and tailpipe fell off on the highway. Then I felt better, we had diddled up even.

The muffler was no big deal either – five bucks for a Canadian Tire flex hose and I straight-piped on down the road.

I went to work for a large engineering company. Trenton AFB was being expanded with more hangers and new runway approaches. I would be a quality control inspector, giving me much more power than I either wanted or deserved. The pay was excellent, even if there was a deduction or two.

I started out at a scale watching an old guy weigh trucks. I worked Monday to Friday and it was boring. The old guy was nice, but seemed to have led a boring life. On my part I didn't want to upstage him with tales of my summer.

But it was mid-October, the weather was great and the Golden Hawks were out and about.

The scale shack sat a couple of hundred yards off to the side of the main runway. The RCAF Aerobatic Team was based in Trenton and they flew gold F-86 Sabre Jets – the Golden Hawks. Every day we got a demonstration of takeoffs and landings – singles, pairs, and often in full formation.

When the truck haul wrapped up I moved to the construction site and tested cement until early December. There was little snow as yet, but freeze-up was upon us and soon the cement work would cease for the winter months. In mid-December I packed my stuff, said goodbye to Patsy and headed home to the farm.

I left early on a Saturday morning – It would be a long, non-stop trip through Northern Michigan, Wisconsin and Minnesota. It was a clear day, getting colder as I neared Sudbury, but the Plymouth had a good heater and I drove coatless, snug as a bug in a coupe. West of Sudbury towards Sault Ste. Marie, the voltage regulator packed it in. The ammeter needle was glued to the far right, so I stopped, lifted the hood and tried the old tap-the-regulator-with-a-screwdriver fix. No luck, so I removed the cover and separated the contact points, thinking they may have just been stuck together. When I fired up, the needle went to the far right again, stuck to the pin.

It was Saturday, service stations were open but no mechanics were on duty, and even if I could find an open garage, I doubted they would have a voltage regulator on the shelf. I didn't want to hole up in a motel room until Monday, so I had no choice but to forge onward. I turned on every switch the Plymouth had – which wasn't many – headlights on high beams, heater on high and although the radio only hummed I turned it on also. Even the interior dome light and the electric windshield wipers joined the fight. This brought the needle back off the pin a bit, and I figured that now the battery would not boil – I just hoped the generator could handle it. I drove on into Northern Michigan getting flack from nearly every oncoming vehicle. In 1959 no one drove with their lights on in the daytime – only this hick in an old Plymouth, so headlights were flashed, saying, "Wake up, dummy!"

The winter solstice was only a couple of days away and even on clear days darkness arrived early, so when I hit Wisconsin the sun went down and the generator burned out! I had already lost a lot of time – lower rpms had also helped keep the needle slightly under control. Now I could speed up, which I did. I had to get home before the battery failed.

What a deal – I shut off everything but the headlights and heater. The heater stayed on low speed, just enough to keep the windshield defrosted, so I donned my winter coat.

At Duluth the headlights were weakening, so the heater was retired. I now had to drive with the vent window cracked, equalizing inside and outside temperatures to prevent the windshield from frosting up. I dug out my toque, mittens and wool socks. Brrr - it's cold inside!

After midnight, heading north on Minnesota #53, the traffic started to thin out. With a full moon reflecting on pure white snow, I could now run with no headlights. Lights on in the daytime – lights out at night – someone was unclear on the concept.

I cleared customs well after midnight with six miles to go. No headlights at all now. They were useless anyway and there was zilch for traffic. I pulled into the farmyard at 4 am and the Plymouth died before I turned off the ignition.

The second post-grad course was completed and I had made it by the skin of my Fiddle Foot toes.

Me with nephew Steve on the re-regulated Plymouth. (Note - "baby moon" hubcaps.)

On January 2, 1960 I left for Winnipeg with a re-regulated Plymouth to look for work. I bunked at my sister's for two weeks and sold the Plymouth for $200.

My last post-grad semester proved to be a throw-away course – something NOT to pad my resume with. What it did teach me was that while there may be no such thing as a poor job, some are certainly better than others.

It was at an electrical wholesaler close to downtown Winnipeg. I would be trained as a shipper/receiver, so I moved into a nearby boarding house and I shipped, and I received, and I received, and I shipped. Even the weekends were boring – there was little cash left over after paying the landlady. There was always the chance of advancement, perhaps on to the order desk, then maybe a few years later to a desk in the front office, requiring the dreaded suit and tie!

So I called the Jet Moving and Storage guy, fully intending to ask him for a job. He was happy to hear from me and we played catch-up. He said he was buying into the company but things were slow this winter – he was the only full-time employee. I said goodbye and good wishes and never did brace him for a job that I knew was not there.

In April, with a boring Winnipeg Easter Weekend on tap, and with a ride available, I went home for three days. Saturday evening, I stopped at a bar in Fort Frances. Whether it was pre-destination, or a withdrawal from the good-luck bank account, a chance meeting that night changed my life forevermore.

At the bar I ran into Ralph Johnson, a nice guy I had known in high school. He was working for Canadian Nickel at Moak Lake, Manitoba, and had taken an Easter break. I told him I had a nothing job and he suggested that I should contact Canadian Nickel. I went back to Winnipeg with an address in my pocket and wrote a short letter to my contact person in Moak. As far as job applications go, mine was pretty sketchy. I left out my Fiddle-Foot resume, saying only that I was farm-raised and had some survey experience. Basically – I wanted a job should one be available.

Moak Lake didn't answer. In June I quit shipping and receiving and went to Atikokan to look for work, my April letter to Moak Lake forgotten. I was in Atikokan for two days when my parents contacted me.

Canadian Nickel had called! They wanted me to proceed to Thompson Manitoba soonest!!!

THE END

Epilogue

Thus the "I Call Myself a Prospector" trilogy is now complete. Book Three now leads back to Book One – a closed circle. It's enough to make your head spin.

Writing has kept my brother and I busy for five or six years, with occasional interruptions to deal with ordinary life. It has been an enjoyable time-filler and our one big plus while working on Book Three is that we have had a chance to examine our own family history in depth.

But – the final audit shows up on the ledger reading like a January weather forecast – income minus expenses equals Below Zero. That's Ok – prospectors, expect this.

And prospectors we are still. Frank is sampling claystone occurrences in the Big Muddy while I, as acting CFO for the Sandstorm Grubstake, work the phone centre.As per usual, it's Minus Zero dollars out there.

So don't you worry none – we won't live long enough to write another book, and even if we do have many years ahead, we are too tuckered out to get into any more shenanigans – or are we?!!

One Final Table Scrap: As Told to Me by (Who else?) Sam Duggan

(This is a Scotty MacDonald story. You remember Scotty from Book Two? He was last seen flying over a '57 Ford in the Red Lake Inn parking lot, having been body tossed by the hi-grade cop.)

In the winter of '69 The Uchi Lake rush was on. Scotty and his staking partner were getting up in years, but they decided to grab a claim block. They would work together – "safety in numbers."

Partner broke trail while Scotty followed, blazing the claim line. Partner had a mild ticker problem. He told Scotty that if he dropped, Scotty was to take a heart pill from Partner's shirt pocket and slip it under his (Partner's) tongue.

Partner dropped. Scotty gave him a prod with a snowshoe, saw Partner was still breathing, muttered, "Lazy old bastard!" and continued on down the claim line. Partner came around, dug in his own pocket for the pill and caught up to Scotty, who was cutting a corner post.

Here's where I expected Sam would tell me that they had a dust-up, but not those two – they'd known each other forever. Partner merely said, "It's your turn to break trail."

So – old prospectors never die; they just rest a while on the snowshoe trail.

Picture courtesy Cypress Development Corp. and Bob Marvin

Made in the USA
Lexington, KY
10 July 2019